Industry 4.0 in Small and Medium-Sized Enterprises (SMEs)

Industry 4.0 in Small and Medium-Sized Enterprises (SMEs)

Opportunities, Challenges, and Solutions

Edited by
Ketan Kotecha, Satish Kumar,
Arunkumar Bongale, and R. Suresh

CRC Press is an imprint of the
Taylor & Francis Group, an informa business

First edition published 2022
by CRC Press
6000 Broken Sound Parkway NW, Suite 300, Boca Raton, FL 33487-2742

and by CRC Press
2 Park Square, Milton Park, Abingdon, Oxon, OX14 4RN

Library of Congress Cataloging-in-Publication Data

Names: Kotecha, Ketan, editor. | Kumar, Satish (Professor of mechanical engineering), editor. | Bongale, Arunkumar, editor. | Suresh, R. (Professor of mechanical engineering), editor.
Title: Industry 4.0 in small and medium-sized enterprises (SMEs) : opportunities, challenges, and solutions / edited by Ketan Kotecha, Satish Kumar, Arunkumar Bongale, and R. Suresh.
Description: First edition. | Boca Raton : CRC Press, 2022. | Includes bibliographical references and index.
Identifiers: LCCN 2021045814 | ISBN 9781032061313 (hardback) | ISBN 9781032061337 (paperback) | ISBN 9781003200857 (ebook)
Subjects: LCSH: Industry 4.0--India. | Small business--Technological innovations--India.
Classification: LCC T59.6 .I36 2022 | DDC 658.4/03802855745--dc23/eng/20211117
LC record available at https://lccn.loc.gov/2021045814

ISBN: 9781032061313 (hbk)
ISBN: 9781032061337 (pbk)
ISBN: 9781003200857 (ebk)

DOI: 10.1201/9781003200857

Typeset in Times
by Deanta Global Publishing Services, Chennai, India

Contents

Preface

The era of industrial automation has lasted several decades, and now it is time for the fourth Industrial Revolution – also known as Industry 4.0. Large-scale industries worldwide have embraced the Industry 4.0 concepts or are in the implementation phase and are working rigorously to introduce the related enabling technologies. But the same is not the case with micro-, small-, and medium-scale enterprises (SMEs).

Although SMEs play a major role in developing countries such as India and other Asian countries, they are unable to investigate the potential of Industry 4.0. One of the prominent reasons for this is the scarcity of Industry 4.0-skilled human resources and limited financial resources. In addition, the SME sector also experiences other hurdles such as insufficient business support, lack of the right competencies, non-availability of digital standards, online threats, cybersecurity issues, inadequate support from government authorities, and finally, incompetent academic institutions that are unable to impart Industry 4.0 skills to the workforce.

This book will serve as a base to facilitate the smooth adoption of Industry 4.0 technologies by the SME sector. Its main focus is on the four broader areas of Industry 4.0: recent advances in Industry 4.0 and smart manufacturing; challenges and opportunities for SMEs to implement Industry 4.0 practices; enhancing the role of academia in upskilling; and digital transformation in Industry 4.0. Each of the areas will further delve into the key concepts of predictive maintenance, the advent of digital factories and digital twin, and additive manufacturing and machining for sustainable development. The book will also discuss the challenges faced by Industry 4.0, such as the need to adopt effective security and privacy policies in the smart manufacturing set-up, and the impending impact of Industry 4.0 on lean production systems. Finally, the book will present the pivotal role of engineering education and curriculum for the skills development of Gen Z engineers embarking on their journey towards Industry 4.0.

Seeking to promote research related to these new trends and developments in Industry 4.0 and its implementation in small- and medium-scale enterprises, this book is divided into 12 chapters. Chapter 1 introduces Industry 4.0 and related technologies such as smart factories, cyber-physical production systems, intelligent manufacturing, and digital manufacturing. In addition, issues related to technology implementation, such as security issues, data availability and storage, and the requirement for high-speed networks are also discussed in this chapter.

Chapter 2 discusses issues related to SMEs and their growth in the global competitive market. This chapter explores the reasons such as low adherence between the governing dimensions of their bushiness, customers, processes, employees, and finances in the context of Indian SMEs and prescribes a way forward to achieve Industry 4.0 readiness. The chapter also includes a survey of 200 firms showing that the performance of SMEs was inconsistent, resulting in defects, scrap, and rework. Finally, the chapter concludes that Indian SMEs are less automated, and the processes impact employee motivation directly and, in turn, customer satisfaction. Thus, Industry 4.0 can increase the consistency and effectiveness of Indian SMEs.

Chapter 3 discusses the construction industry and its readiness to adopt Industry 4.0. The chapter describes the four pillars of the construction industry as visualisation, integration, communication, and intelligence. The issues of data generation, tapping, and data management in construction projects, construction site safety data, classification and codification capability, and information modelling have also been discussed in detail. The systematic adoption of Industry 4.0 principles for construction site safety management is also discussed in this chapter.

The concept of predictive maintenance of production machines is discussed in Chapter 4. Machinery fault detection under the umbrella of predictive maintenance helps monitor machinery operation via integrated sensors, generates alerts on observing noticeable deviations in sensor readings, and predicts machine failure ahead of time. This chapter delves into the concept of artificial intelligence (AI)-enabled machinery fault detection and provides an in-depth review of traditional and modern approaches used in fault detection. It further explores the recent advancements in the sensors used and extracted for fault detection in case studies. This work would be a resourceful literature for researchers and practitioners planning to explore the impact of artificial intelligence in smart manufacturing.

The challenges in the agricultural sector to adopting Industry 4.0 principles, such as integrating wireless sensor networks to monitor and control crop growth and crop harvesting using mobile robots, are addressed in Chapter 5. In addition, this chapter discusses recent trends in combining sensor and actuator networks and machine learning techniques such as reinforcement learning and kernel-based methods for spatiotemporal modelling and control purposes in smart farming applications. Illustrative examples and the advantages of their use in agricultural SMEs are also discussed.

Chapter 6 presents a data-driven, digital twin of a complex laboratory gas turbine engine (GTE) using a deep learning framework. A deep neural network comprising long short-term memory (LSTM) cells is developed that maps the three inputs with seven output parameters of a laboratory GTE to model the dynamic response. The prediction performance of LSTM networks is compared with a conventional non-linear autoregressive network and is found to be quite efficient in predicting the dynamic response of GTE parameters. Furthermore, the prediction response shows that the LSTM network–based digital twin outperforms the mathematical model.

Chapter 7 is dedicated to discussing additive manufacturing and 3D printing technology in Industry 4.0 using a case study of prostheses development. Additive manufacturing is one of the emerging techniques for producing equipment for medical applications. The focus of this chapter is on a lower limb prosthesis, namely the prosthetic socket. The proposed approach is based on 3D scanning of a patient's residual limb and developing the prosthetic socket using polymer additive manufacturing (AM) utilising the fused deposition modelling (FDM) technique. In addition, future challenges and opportunities are presented.

Chapter 8 discusses a case study on challenges in the printing sector. This chapter aims to identify the technological gaps in the Mysore Printers Cluster and provide recommendations to bridge the gap to create an attractive opportunity in the Industry 4.0 market. Based on the various surveys, inputs, recommendations, and suggestions provided for this project, an action plan is provided that emphasises various training modules with topics, and a brief content is proposed for future training requirements at the Mysore Printers Cluster, to enhance the knowledge and cognitive skills of the employees for market sustainability. The action plan is achieved by employing domain experts to reach the grass root level in the targeted audience's minds, which would lead to an attractive opportunity in the Industry 4.0 market.

Chapter 9 reviews and discusses recent trends in intelligent machining. The chapter is divided into three interrelated sections, i.e. machine tools, cutting tools, and techniques. The machine tools section is focused on intelligent machine tools used in intelligent machining. The cutting tools mainly refer to smart cutting tools and their real-time condition monitoring in intelligent machining. The techniques section focuses on soft computing techniques and assisted optimisation techniques used in intelligent machining. Finally, the chapter summarises development trends in intelligent machine tools, cutting tools, and optimisation techniques.

Chapter 10 describes various initiatives in India to strengthen the SME sector and enable it to adopt Industry 4.0 principles. Several initiatives, such as the Digital India campaign (2015), the Make in India initiative (2014), Start-up India (2015), and Atmanirbhar Bharat (2020) have been introduced to boost the digital economy of India. This chapter begins by delineating how government schemes/initiatives helped boost the country's digital infrastructure and aided the incubation of digital start-ups that became market leaders. For this purpose, the case study method was utilised to collect information on government initiatives and the digital economy of the country, and is presented in this chapter.

The skills requirement of the workforce and the current level of skills are significant concerns for industries, including SMEs, to transform from traditional manufacturing processes to the new Industry 4.0 revolution. Chapter 11 highlights the necessary skills required by the present workforce for Industry 4.0 and the challenges faced by industries in meeting the need for skilled human resources.

Chapter 12 discusses the changing role of academics from the perspective of educational transformation in this era and beyond. First, the chapter introduces the Education 4.0 concept and discusses the evolution from the traditional concept to the emerging one. The chapter then discusses the new era's key skills and competencies, including soft skills, challenges, and solutions that the digital transformation in education has brought to higher education institutions (HEIs). Finally, future-ready education ecosystems, together with their active participants, are discussed.

The editors acknowledge their gratitude to CRC Press/Taylor & Francis Group for this opportunity and their professional support. Finally, we would like to thank all chapter authors for their interest and availability to work on this project.

Dr Ketan Kotecha
Dr Satish Kumar
Dr Arunkumar Bongale
Dr R. Suresh

Editors

Ketan Kotecha gained his doctorate degree from the Indian Institute of Technology (IIT) Bombay in 2003. Dr Kotecha's core areas of research interest are artificial intelligence, machine learning, and deep learning along with computer algorithms and machine learning, higher-order thinking skills, critical thinking, and ethics and value. He was principal of GH Patel College of Engineering and Technology, and then served as director of Nirma University. He was also vice chancellor of Parul University. Currently, Dr Kotecha is working as the dean of the Faculty of Engineering and director of the Symbiosis Institute of Technology, Symbiosis International (Deemed University). He is head of the Symbiosis Centre for Applied Artificial Intelligence (SCAAI) and chief executive officer of the Symbiosis Centre for Entrepreneurship and Innovation. Dr Kotecha is a member of the National Advisory Council for the Confederation of Indian Industry's (CII) Engineering and Management Curriculum Restructuring Task Force. He is a technical advisor to a team for the bus rapid transit system (BRTS) implementation in Ahmedabad. He has also been appointed as an independent director of Gujarat Informatics by the Government of Gujarat. Dr Kotecha has a total of 45 publications in Scopus-indexed journals to his credit. He has guided 13 PhD scholars and 6 students are taking his guidance for PhD.

Satish Kumar is an associate professor in the Department of Mechanical Engineering and is in charge of the Advanced Manufacturing Technology Laboratory at the Symbiosis Institute of Technology, Symbiosis International (Deemed University), Pune, India. He is also in charge of the PhD programme in the Faculty of Engineering (FoE), Symbiosis International (Deemed) University. He completed his master's degree (MTech-Tool Engineering) in 2013 and his doctoral degree (PhD) in 2020 from Visvesvaraya Technological University, Belgaum, Karnataka, India. He has over eight years of experience in teaching, research, and industry. His areas of research interest include smart manufacturing, digital twin, condition monitoring, composites, cryogenic treatment, additive manufacturing, and hard materials machining. He has authored more than 32 international/national journal and conference publications. He has filed Indian patents based on his current research projects and he is also a corporate member of the Institution of Engineers. He is a research supervisor to PhD research scholars and M.Tech students working in the domain of predictive maintenance, manufacturing, and Industry 4.0.

Arunkumar Bongale is an professor in the Department of Mechanical Engineering, and is in charge of the workshop department at the Symbiosis Institute of Technology, Symbiosis International (Deemed University), Pune, India. He has a bachelor's degree in mechanical engineering, a master's degree in computer-integrated manufacturing, and a PhD in mechanical engineering. He has more than 15 years of teaching and research experience. His research interests are smart manufacturing, robotics and automation, material science, heat treatment, composite materials, and composite coating technology. He has authored more than 25 international/national journal and conference publications in the relevant areas of manufacturing technology, characterisation of materials, composite manufacturing, and evaluation. His ongoing projects include multisensor-based real-time monitoring and control of friction stir processing methods; electromagnetic field–assisted electric discharge machining – an Industry 4.0 approach; and the development of plasma arc oxidation methods for industrial applications. Dr Bongale has filed two Indian patents based on his current research projects. He is a corporate member of the Institution of Engineers and a member of the Society of Automotive Engineers.

R. Suresh is an professor in the Department of Mechanical and Manufacturing Engineering at M.S. Ramaiah University of Applied Sciences, Bangalore, India. He gained his master's degree (M.Tech) and doctoral degree (PhD) in 2002 and 2013, respectively, from Kuvempu University, Karnataka, India. He has 24 years' experience in teaching, research, and industries. His areas of research interest include composites, surface coating, additive manufacturing, hard materials machining, numerical modelling, and experimental analysis of advanced materials. He has authored more than 60 international/national journal and conference publications leading to h-index. He has supervised PhD, undergraduate, and postgraduate students. He is a member of MISTE, SAE India, and other professional societies. He has received research funding from agencies such as DST, AICTE, and KCTU – Government of Karnataka, and he has worked on many industrial projects.

Contributors

Sıtkı Akıncıoğlu
Department of Machine Design and Construction
Duzce University
Duzce, Turkey

Varsha Bagodi
City University of London
London, UK

Virupaxi Bagodi
Government Engineering College
Talakal, India

Arunkumar Bongale
Symbiosis International (Deemed University)
Pune, India

G. Devakumar
Innovation and Entrepreneurship Development Research Centre
M.S. Ramaiah University of Applied Sciences
Bangalore, India

Ajit Dhanawade
College of Engineering and Research
Pune, India

Shalaka Hire
Symbiosis Institute of Technology
Symbiosis International (Deemed University)
Pune, India

Priya Jadhav
Symbiosis Institute of Technology
Symbiosis International (Deemed University)
Pune, India

Ajith G. Joshi
Department of Mechanical Engineering
Canara Engineering College
Bantwal, India

Pooja Kamat
MIT-ADT University
Pune, India
and
Symbiosis Institute of Technology
Pune, India

F. Karaferye
Kutahya Dumlupinar University
Kutahya, Turkey

Satish Kumar
Symbiosis International (Deemed University)
Pune, India

M. Laad
Department of Applied Sciences
Symbiosis Institute of Technology
Pune, India

Arnab Maity
Indian Institute of Technology Bombay
Mumbai, India

M. C. Murugesh
Department of Mechanical and Manufacturing Engineering
M.S. Ramaiah University of Applied Science
Bangalore, India

Vishal Naranje
Amity University
Dubai International Academic City
Dubai, UAE

P. S. V. Nataraj
Indian Institute of Technology Bombay
Mumbai, India

Sayak Pal
School of Media and Communication
Adamas University
Kolkata, India

M. Renedo
M & M Profuture Training
Barcelona, Spain

Kirti Ruikar
Indian Institute of Foreign Trade
Kolkata, India

Sayali Sandbhor
Indian Institute of Foreign Trade
Kolkata, India

Richa Singh
Indian Institute of Technology Bombay
Mumbai, India

Deepankar Sinha
Indian Institute of Foreign Trade
Kolkata, India

Rekha Sugandhi
MIT School of Engineering
MIT-ADT University
Pune, India

R. Suresh
Department of Mechanical and Manufacturing Engineering
M.S. Ramaiah University of Applied Science
Bangalore, India

T. A. Tamba
Parahyangan Catholic University
Bandung, Indonesia

Nitesh Tripathi
Department of Journalism
Adamas University
Kolkata, India

Seema Wazarkar
Symbiosis Institute of Technology
Symbiosis International University
Pune, India

1 Industry 4.0: An Introduction in the Context of SMEs

Priya Jadhav, Satish Kumar, and Arunkumar Bongale

CONTENTS

DOI: 10.1201/9781003200857-1

1.1 INTRODUCTION

The focus of the first Industrial Revolution was primarily on increasing mechanical productivity. The second Industrial Revolution in the 19th century was aided by the use of electricity to increase production speed while keeping costs low. The use of memory-driven computers and some automation led to the third revolution. With the introduction of 3D printing technology and intelligent, self-monitoring logistics systems, the Industrial Revolution has progressed significantly.

We are currently in the fourth Industrial Revolution with the digitisation of production systems improving with various artificial intelligence (AI) techniques such as machine learning and big data approaches. The various supporting smart devices, the industrial internet of things (IIoT), play a crucial role in integrating hardware or machines with the data and software systems. Hence, modern manufacturing systems are considered smart systems with real-time decision-making based on data analysis algorithms. Together with the use of robots and the automation of subprocesses, difficult tasks are made simple. Hence, Industry 4.0 makes manufacturing systems eco-friendly, sustainable, and economic.

This chapter focuses on the applications, methodologies, and associated data, communication, and security problems, as well as the opportunities that AI and big data bring to Industry 4.0. We examine a variety of AI and big data methodologies in depth, as well as the Industry 4.0 applications that have profited from AI and big data. The chapter also highlights and examines major technological, data-related, and security risks and problems that come with successful AI and big data deployment in Industry 4.0.

This chapter covers the following key concepts in Industry 4.0:

- Digital manufacturing
- Smart factories
- Challenges in Industry 4.0 adaptation
- Opportunities in implementation

1.2 DIGITAL MANUFACTURING

Over time, generalised manufacturing systems have become customer requirement oriented, including the manufacture of parts and inventory management. Developing excellence in the field by creating a brand name and maintaining network systems with other manufacturing units is important for collaboration. This can be achieved using flexible manufacturing systems (FMS) or other techniques for adopting a greater number of changes in the schedules or set-ups of assembly lines. Still more high-end goals where automation is an integral part of the production system need more intelligence systems. Hence, manufacturing systems need to be more competitive to compete in the market (Bousdekis, Lepenioti, and Apostolou 2021; Pech and Vrchota 2021).

Product design, production, and maintenance management have been part of the traditional production method. Changes in design are the key step to improving quality and reducing costs in the initial stages of production. Supporting tools such as computer-aided design (CAD) and computer-aided manufacturing (CAM) benefit from virtual simulations that deliver faster and more reliable results. These simulations are an integral part of the design, quickly validating the shop floor management data for larger assembly lines.

In comparison to previous mass production techniques, lean manufacturing involves managing product development, production systems, and customers with less effort, less time, and better quality. The term "agile manufacturing" refers to the process of bringing a new product to market through modifying production systems. This fosters a culture of confronting issues and adapting to new approaches as market conditions dictate. On the other side, sustainable manufacturing integrates society, the environment, and the economy from an operational perspective, focusing on resource use, worker engagement, and community development (Chong, Ramakrishna, and Singh 2017). When these three pillars of lean, agile, and sustainability (Figure 1.1) are examined as a single system, lean emphasises a system's

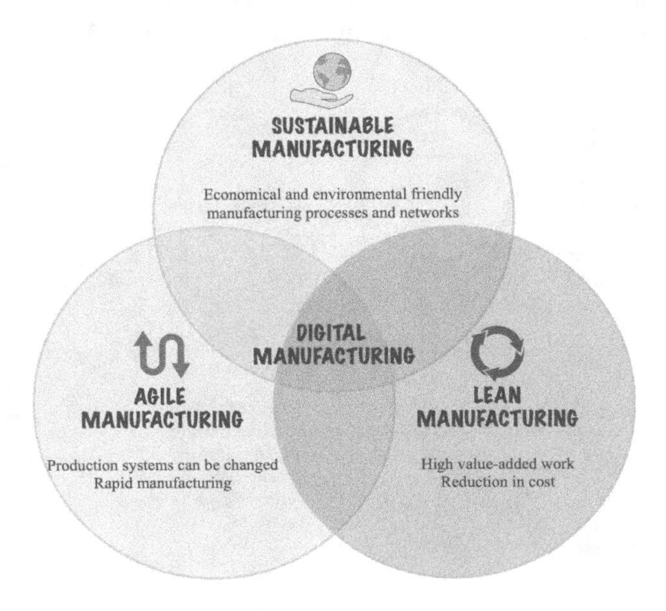

FIGURE 1.1 Foundations of digital manufacturing.

stability, which can be referred to as autonomy, while agility adds the ability to adapt to new situations, focusing more on collaboration (Nylund and Andersson 2001).

Changes can also come from spontaneous suggestions that are perceived as improving the system's competence. If the process has not yet been built, the current system must be analysed in order to create digital information and knowledge of what is now available. The new needs for the future model are formed by a synthesis of the existing system and possible adjustments. The solution principles are formed by a mixture of plausible new options and current capabilities. The end product is a set of integrated structures and abstract and conceptual formulations, which would include the future system's goals and basic attributes. Possible technologies can be researched as the descriptions are more thoroughly examined, resulting in different solutions. The solution options can be modelled as virtual entities with their own operating rules, motion, and behaviour in addition to their digital description. A basic simulation model is created by combining existing and new virtual components.

Digital manufacturing relates to monitoring and controlling the production system using different sensors and actuators, and analysing the data obtained from them for monitoring and controlling purposes. Digital manufacturing starts with converting the digital data into measurable figures to perform an analysis using the various algorithms for monitoring. The digital systems can decode the data in the form of binary data and images as well as sound. The approach for the development of digital manufacturing is to develop computer-integrated programs that can improve the quality and quantity of production systems (Figure 1.2).

The internet is a fusion of data and services between physical and digital machines. The drivers of innovation are the new factory model. One has to promote sustainable development, the efficient production of system resources, innovation, and a successful economy (Chong, Ramakrishna, and Singh 2017). A manufacturing

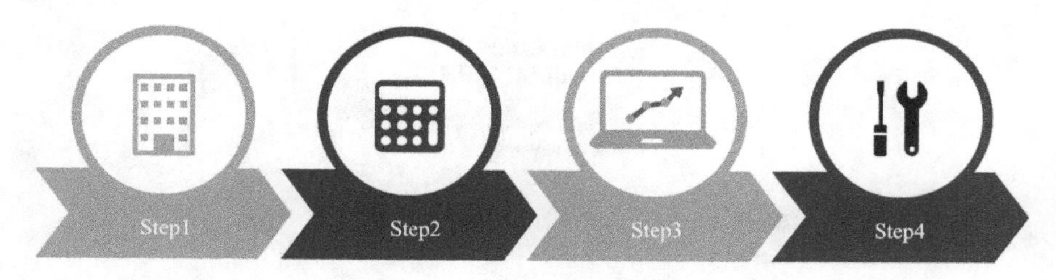

FIGURE 1.2 Steps in digital manufacturing.

system's operation can be seen in three time dimensions: past, present, and future. The past depicts what has occurred, and it might be considered the system's digital memory. The present time dimension, or what is going on right now, is used to run the current system by monitoring its state and comparing it to the desired state. The future dimension allows for future manufacturing planning. The data collected from a system's activities as they occurred is displayed in the past. It can be used to investigate earlier manufacturing activity in order to determine what went wrong and why. The system can learn from its past and avert unpleasant events in the future by determining the root causes of problems. Regulations on the autonomy of manufacturing entities, as well as their collaboration, can be improved, and new rules can be devised. The term "present" refers to a time period in the near future when no major changes are expected.

1.3 COMPONENTS OF INDUSTRY 4.0

The machines, devices, and connecting interfaces have different layers for components of Industry 4.0. These components are grouped based on different principles.

On the shop floor, the controllers and smart IoT devices operate together with computerised numerical control (CNC) and other digital machinery. These devices are used to store the data that these machines generate. In general, connected devices such as Wi-Fi, Bluetooth, and Zigbee allow for real-time data monitoring. The information from these devices is kept up to date and aids in supply chain management. The linkages of the devices are also crucial in comprehending data analysis. This analytic study serves as a foundation for machine learning algorithms, thus it must be as error-free as possible (Sufian et al. 2021). The interaction of various IIoT devices for data storage, analysis, and application end status checking is depicted in Figure 1.3. The applications display data to end users in the form of chatbots or augmented reality (AR), mostly for data visualisation. For data connectivity, the following data tools are needed.

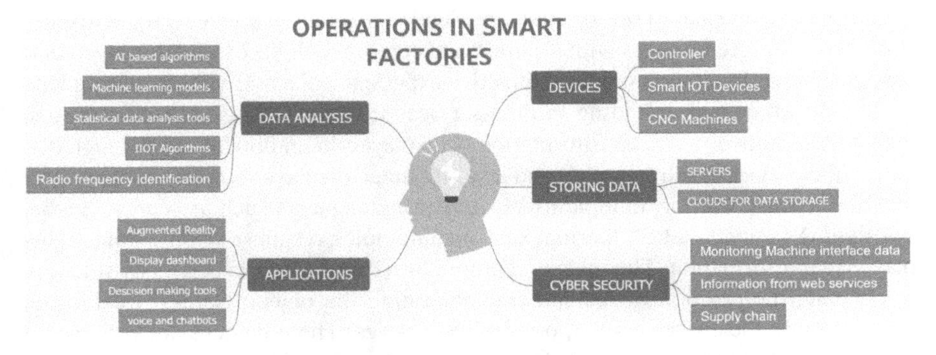

FIGURE 1.3 Layers of digital device interaction and their structure Industry 4.0.

1.3.1 Connectivity of Sensors

Intelligent machinery, sensors, and control systems that monitor a strong framework of manufacturing operations make up smart factories. These advancements have changed not only factory floor architecture, boosting consistent and accurate collaboration between machines, but also machine needs, increasing the demand for trustworthy sensors. Sensors and devices connect industrial equipment, hardware, and software. The integration of various systems results in a system that is easier to operate, more convenient, and less faulty (Kalsoom et al. 2020). Sensors can work passively or actively depending on the responses generated at the required location. The signals can be physical such as light illuminated or temperature sensors. Smart factories use numerous sensors starting from positioning to sensing the component features.

1.3.2 Data Storage and Cloud

Data is acquired from a variety of sources in smart industries, including smart machines, various plants, and the products made in those facilities. Industrial big data can be utilised alone or in conjunction with other platforms such as mobile/edge devices, cloud services, and on-premise systems. Cloud services have become an essential component of digital production systems. They are responsible for the storage and protection of large amounts of data at various levels of operation.

1.3.3 Augmented Reality

The data from 3D models at the development and manufacturing stages is used to produce a process support tool that streamlines the process and lowers the error rate.

AR is a set of technologies that employ an electronic device to observe a real-world physical environment that is merged with virtual aspects, either directly or indirectly.

A component for capturing photographs, the virtual information will be visible on a display. Retinal projection is also rising in popularity; however, it is not widely used.

The virtual and actual information are blended together in a video-mixed display. Previously captured images with a camera are combined digitally and displayed on a screen. This technology provides interactive experience of a real-world environment.

The drawbacks of providing limited services include decreased resolutions and vision. Alternatively, virtual information is displayed in an optical see-through display. Other devices such as spatial and head-mounted displays show the collaborative task for a single user or multiple users. Activating elements such as sensors, global positioning systems (GPS), thermal sensors, and quick response (QR) scanners display visual information. The images captured by a camera are processed by software to estimate the position of the camera concerning the object. Natural tracking of features gives the characteristic points of the images. These devices are very useful at critical or hidden positions where human presence is difficult to monitor (Fraga-lamas, Fernández-caramés, and Senior Member 2018).

1.3.4 Artificial Intelligence for Departments

Artificial intelligence tools are used to improve quality, detect errors, analyse data, and increase productivity. Customer service, maintenance, logistics, and inventory management are all improved as a result. Artificial intelligence algorithms predict market demand by looking for patterns that link, among others, regional, socio-economic, and financial aspects, weather patterns, political status, and consumer behaviour. Manufacturers benefit from this data because it allows them to manage personnel, inventory control, energy consumption, and raw material supply. Machine learning technology and pattern recognition software are at the heart of AI's incursion into industry, and they may eventually lead to reshaping industries in the near future (Figure 1.4).

1.3.5 Cybersecurity

Cybersecurity serves as a barrier between smart machines and the outside world. Cybersecurity can be applied to machines, hardware, and software components to be obliterated by putting in place strong cybersecurity solutions in intelligent industries.

1.3.6 Digital Twin in Smart Factories

To improve business outcomes, a product's life cycle can be virtually monitored with digital twins. This is real-time monitoring of machining data that updates itself

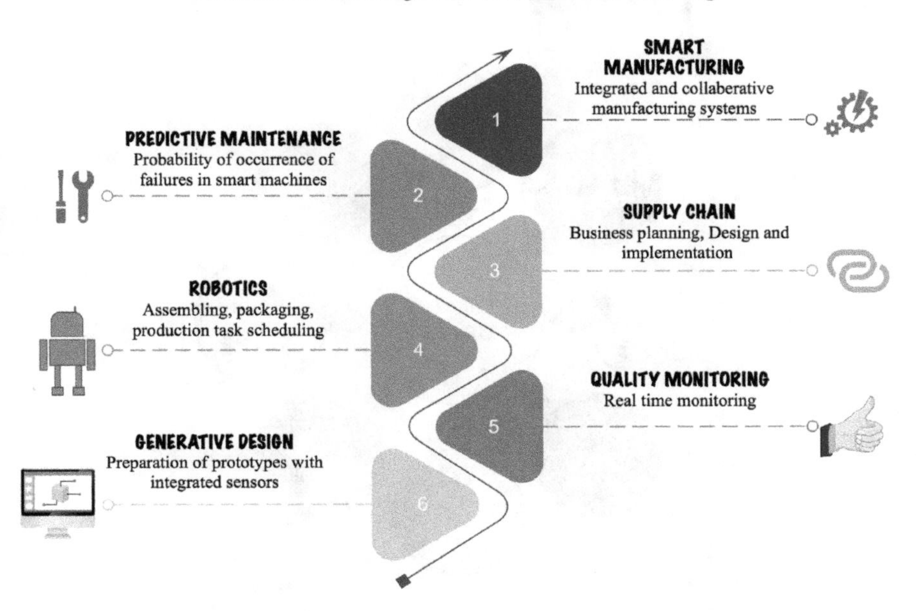

FIGURE 1.4 Application areas of AI in manufacturing.

using historical data via machine learning algorithms. It enhances defect detection, machine breakdowns, and opportunity forecasts (Alcácer and Cruz-machado 2019). Furthermore, as models and physical processes co-evolve, fresh data is generated by the models. The models are used to communicate. A Digital Twin technique for recording and analysing the activities of equipment or systems to estimate their future condition based on their current condition data in real time, data from the past, experience, and expertise (Tao et al. 2017).

1.4 IIOT IN INDUSTRY 4.0

The IIoT is a rapidly growing topic in the manufacturing industry since an increasing number of smart devices and sensors generate significant amounts of data. IIoT applications include device connectivity, cloud storage, and algorithm processing. For the end user, this data should be easily accessible, intelligible, and analysable. Figure 1.5 summarises the key areas handled by the IIoT. It is critical for businesses to comprehend and anticipate the influence of technology and robotic systems, and to plan their operations accordingly. While numerous lessons can be learned from previous automation episodes, artificial intelligence and robotics may have new effects (Alcácer and Cruz-machado 2019).

As the data collected from various devices and sensors is heterogeneous, it is necessary to interpret this data and keep track of various operations on the manufacturing line. The collaborative scheduling of different sets of jobs eliminates unnecessarily long wait times in unneeded applications. However, they are vulnerable to issues such as data storage limitations and the processing of large amounts of data.

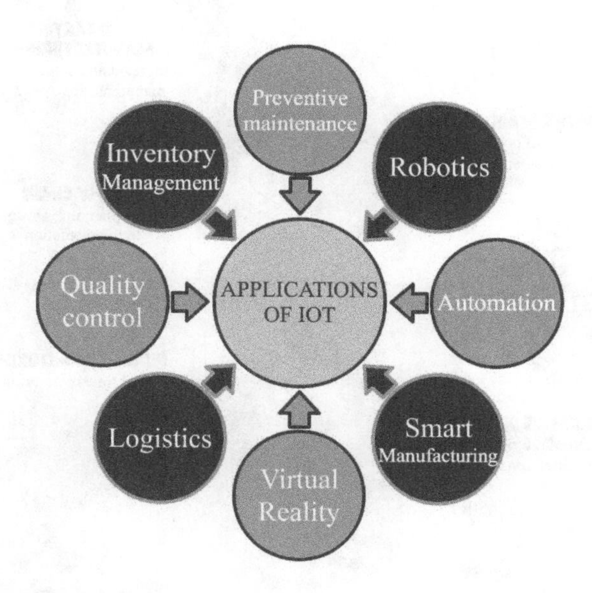

FIGURE 1.5 IIoT applications in Industry 4.0.

1.5 SMART FACTORIES

The new trend in manufacturing is the demand for a customer-oriented product, which raises the requirement for flexible manufacturing systems. As the life cycle of products is getting shorter, companies need to adopt new technologies to complete the demand in time.

Smart manufacturing can be divided into horizontal integration, vertical integration, and end-to-end integration. Horizontal integration looks after collaboration with other companies while vertical integration synchronises between the different subsystems of the company. The third type of integration, end to end, refers to the combination of the design, customer requirement, and flexibility in production schedules. Effective working of all of these ensures a smooth and efficient production process. Smart manufacturing consists of multiple resource management, the dynamic routing of a process as per requirement, and the instant interconnection between subsystems and big data obtained at each step of a process (Alem and Rocha 2021; Mittal et al. 2019). To analyse the data, various smart techniques have been used, a few of which follow.

1.5.1 Deep Learning

Deep learning has its origins in the idea inspired by structure of a human brain. In deep learning the back-propagation approach is used to optimize the weights of the neurons till convergence is not achieved. The use of different IoT devices and data obtained from sensors provides large data for analysis.

In manufacturing, deep learning is used in domains such as quality inspection, fault detection and identification of components and processes, predictive analysis, and condition monitoring (Duan and Yan 2020). The model can be trained using previous data and helps to identify defects using robust algorithms.

1.5.2 Radio-Frequency Identification

This technology is used to monitor the task assigned to each object. The system consists of antennas to send signals, readers to collect signals, information systems, and mobile devices. Each tag contains specific information. In manufacturing industries, radio-frequency identification (RFID) are used to improve the process automation in assembly line. The main advantage of this technology is in mass production where tracking of a process becomes very easy. This information can be used in the feedback system to modify the process flow and improve quality.

1.5.3 Data Mining

Numerous sensors and devices send data in the form of images, process variables, and binary formats. The tools from data mining such as statistical methods, machine learning, and pattern recognition use various images. Data mining is generally used to optimize quality, fault identification, useful life prediction of product or process. It is widely used in areas where the precision of a process is needed. Fuzzy logic can be used to monitor the quality of an inspection. Advanced machine learning algorithms are more useful in high precision measurement tasks in applications such as machining and ultrasonic welding. Monitoring the quality at the process level helps to improve the overall efficiency of manufacturing.

1.5.4 Cloud Computing

In a detailed analysis, more data is required which can be obtained through data computing. It defines the model for network access to a common data pool of devices such as servers, storages, and networks. The customer-oriented manufacturing system keeps track of customer needs and restructures the resources. Hence, adaptation becomes easy in flexible system types. It benefits not only at the enterprise level but also at the process level.

1.6 CHALLENGES AND OPPORTUNITIES

1.6.1 Implementation Challenges

1.6.1.1 Need for Technology Maturity

Highly complex data classification and storage require a more advanced tool for digitisation.

1.6.1.2 Insufficient Technology to Integrate with Manufacturing Systems

To implement Industry 4.0 into an entire manufacturing process, the vertical and horizontal integration of digital systems is necessary. Various factors such as company policy, funding, and a skilled workforce affect the quality of implementation.

1.6.1.3 Difficulty in the Reconfiguration of Production Systems

Updating existing systems becomes tricky with limited funds and a shortage of space. Conventional manufacturing set-ups are not flexible enough to accommodate changes in the system with minimal effort (Harris et al. 2019).

1.6.2 Challenges for Data

1.6.2.1 Unstructured Format of Data

Efficiency in extracting useful information from various devices and transforming the data into a form suitable for different machines and devices has not yet been achieved. The various formats obtained from the source need to be classified correctly to reframe a model. The hierarchy of the data and multiple formats create complicity in classification.

1.6.2.2 Management of Huge Data for Storage and Processing

More advanced storage technology is needed to analyse huge data individually.

1.6.2.3 Quality of Data

Often, data is obtained from various formats that can be inclusive of reductant and noisy data. This affects the quality of the overall data obtained. The traditional tools used for such data have to be updated to get more accurate data classification. More advanced tools can identify more reliable data.

1.6.3 Skilled Workforce

A workforce with a multidisciplinary knowledge of data analysis, mathematics, and management along with basic engineering knowledge is desired. The open-mindedness of an engineer is much needed to adopt new technology tools for immediate and effective implementation on the worksite. Due to continuous development of the trends and technology followed in Industry 4.0, upgrading of the workforce is an ongoing process (Bajic et al. 2020). Workers will be able to focus on sophisticated and inventive jobs as AI takes over the manufacturing facility and automates mundane and dull human tasks. Automation does not replace human labour, but it does complement it and increases productivity in ways that increase demand for human labour. The decline in the cost of performing mundane operations with computers, which complements more creative and innovative service jobs, can be linked to the increase in high-skill employment (Choudhary et al. 2021). Robots will be fully trained to overcome difficult hurdles or ergonomically problematic positions for the end user. The monotonous jobs in the production process will be carried out by smart, autonomous, and self-regulating production processes, thereby improving human work life quality (Sony 2020).

1.6.4 Security Challenges

1.6.4.1 Sharing Data with Stockholders

Most manufacturing processes require the semi-processing of a product or supplying the semi-finished parts to the original equipment manufacturer (OEM). Hence, data sharing between stockholders is a critical point of the security issue. Often, the technical capability of the supplier to handle or generate data is insufficient.

1.6.4.2 Connectivity Protocols

Real-time communication during the manufacturing process requires secured monitoring and storage. This must be done without obstructing the connectivity protocols to be followed at different support systems. A more flexible and reliable protocol procedure is a requirement for the implementation of Industry 4.0. Insufficient data protection is also one of the major concerns in the data handling of manufacturing units.

1.6.5 Difficulty in Altering the Manufacturing Set-Up

Ineffectively developed production systems are challenges when it comes to converting technological resources. It is critical and costly to assimilate existing systems with digital transformation.

1.6.6 Standardising the Process

Process standardisation is difficult due to the use of different tools and technologies by different manufacturing units. This also creates an inflexible method during the exchange of information between units. Flexibility can only be considered when manufacturing units are using similar kinds of digitisation technology.

1.6.7 Financial Challenges

The requirement for large amounts of funds constrains the manufacturer from implementing new technologies. Such investment does not provide direct profit which may be felt as a perceived risk for the manufacturer.

1.6.8 Implementation Difficulties at Managerial Level

A company needs a team of cross-domain workers who can develop an algorithm to detect issues during a process. It is difficult to synchronise and maintain the data handled as per company policy. Real-time monitoring of the process requires the installation of different IoT devices and integrating with the data extraction system. The lack of manufacturing systems with sufficient AR and VR technology is also a major point when identifying defects in components.

1.6.9 Environmental Challenges

Hazardous gases may be generated while assembling heavy machinery with high energy consumption during set-up, and their prevention is a major challenge in Industry 4.0 implementation. AI has the potential to have a negative impact on the environment, particularly because of its contribution to the increased development and exploitation of modern equipment. Increased manufacturing and the use of technical equipment will have two negative consequences: climate change (increase in pollution) and natural resource degradation (Bajic et al. 2020, Machado et al. 2020).

1.7 OPPORTUNITIES IN IMPLEMENTATION OF INDUSTRY 4.0

Considering the above challenges during the implementation of Industry 4.0, the following measures can be taken during the planning stage of digitisation:

- Integrating the existing manufacturing system to adopt changes during new technology implementation.
- Developing a technology that can be flexible with other methods of digitisation.
- Establishing a cross-functional team that can monitor the implementation as well as the execution of Industry 4.0.
- Designing a more flexible manufacturing system that can be redesigned if required.
- Investing more in equipment such as servers, processing powers, and storage spaces.
- Developing more systems or algorithms for the compression and storage of data would save resources.
- Investing in more reliable data collection devices and sensors will help to improve the quality of the system.

The main focus while implementation during real-time data analysis is to have high-power processing units that can respond quickly. For security reasons, more strong protocols and cybersecurity should be built in. To allow the exchange of information within the company and to other stakeholders, a standardised and uniform process flow needs to be developed.

From a technology point of view, further development of infrastructure and compatible algorithms should be undertaken to make the algorithms more robust. Developing pilot projects to test an analytical model can help to achieve accuracy. More equipment with IoT, cloud computing, cyber-physical systems, and augmented and virtual reality can predict numerous formats and types of data capturing and analysing.

1.8 CONCLUSION

The scope of Industry 4.0 is to coordinate the activities of planning, producing, and managing manufacturing systems efficiently. Individual manufacturing entities and entire systems are reviewed in detail. Initial stage to successful solutions with subsystems of manufacturing as self-governing, yet closely related digital, implicit, and real provides efficient and effective manufacturing activities. Digital manufacturing has the potential to contribute to the effective planning, development, and operation of manufacturing systems. Other advanced Industry 4.0 technologies that enable smart production are described in the chapter. It also identifies common issues connected with digital transformation in manufacturing small and medium-sized enterprises (SMEs), as well as effective solutions for overcoming issues. Further research should concentrate on identifying the most advantageous clusters or pairings of Industry 4.0 technology applications. As an approach, it has the potential to raise awareness about areas where improvements are required.

The addition of various artificial intelligence tools also introduces data threats; the need for more flexible and appropriate algorithms for various processes; high connectivity in the manufacturing set-up; smart devices and infrastructure upgrades; data analysis and interpretation tools; data conversion into a readable format for end users; and real-time monitoring.

More advanced devices and techniques can be used to achieve interconnectivity and real-time monitoring. As a result, Industry 4.0 is progressing towards the next generation of human-free production systems that are totally automated.

REFERENCES

Alcácer, V, and V Cruz-machado. 2019. "Scanning the Industry 4.0 : A Literature Review on Technologies for Manufacturing Systems". *Engineering Science and Technology, An International Journal* 22: 899–919. https://doi.org/10.1016/j.jestch.2019.01.006.

Alemão, Duarte, André D Rocha, and José Barata. 2021. "Smart Manufacturing Scheduling Approaches—Systematic Review and Future Directions." *Applied Sciences* 11 (5): 2186. https://doi.org/10.3390/app11052186.

Bajic, Bojana, Aleksandar Rikalovic, Nikola Suzic, and Vincenzo Piuri. 2021. "Industry 4.0 Implementation Challenges and Opportunities: A Managerial Perspective." *IEEE Systems Journal* 15 (1): 546–59. https://doi.org/10.1109/JSYST.2020.3023041.

Bousdekis, Alexandros, Katerina Lepenioti, Dimitris Apostolou, and Gregoris Mentzas. 2021. "A Review of Data-Driven Decision-Making Methods for Industry 4.0 Maintenance Applications." *Electronics* 10 (7): 828. https://doi.org/10.3390/electronics10070828.

Chong, Li, Seeram Ramakrishna, and Sunpreet Singh. 2018. "A Review of Digital Manufacturing-Based Hybrid Additive Manufacturing Processes." *The International Journal of Advanced Manufacturing Technology* 95 (5): 2281–300. https://doi.org/10.1007/s00170-017-1345-3.

Choudhary, Shivani, Deborah Herdt, Erik Spoor, José F García Molina, Marcel Nachtmann, and Matthias Rädle. 2021. "Incremental Learning in Modelling Process Analysis Technology (PAT)—An Important Tool in the Measuring and Control Circuit on the Way to the Smart Factory." *Sensors* 21 (9): 3144. https://doi.org/10.3390/s21093144.

Duan, Gui-Jiang, and Xin Yan. 2020. "A Real-Time Quality Control System Based on Manufacturing Process Data." *IEEE Access* 8: 208506–17. https://doi.org/10.1109/ACCESS.2020.3038394.

Fraga-Lamas, Paula, Tiago M FernáNdez-CaraméS, ÓScar Blanco-Novoa, and Miguel A Vilar-Montesinos. 2018. "A Review on Industrial Augmented Reality Systems for the Industry 4.0 Shipyard." *IEEE Access* 6: 13358–75. https://doi.org/10.1109/ACCESS.2018.2808326.

Harris, Gregory, Ashley Yarbrough, Daniel Abernathy, and Chris Peters. 2019. "Manufacturing Readiness for Digital Manufacturing". *Manufacturing Letters* 22: 16–18. https://doi.org/10.1016/j.mfglet.2019.10.002.

Kalsoom, Tahera, Naeem Ramzan, Shehzad Ahmed, and Masood Ur-Rehman. 2020. "Advances in Sensor Technologies in the Era of Smart Factory and Industry 4.0." *Sensors* 20 (23): 6783. https://doi.org/10.3390/s20236783.

Machado, Carla Gonçalves, Mats Peter Winroth, and Elias Hans Dener Ribeiro da Silva. 2020. "Sustainable Manufacturing in Industry 4.0: An Emerging Research Agenda." *International Journal of Production Research* 58 (5): 1462–84. https://doi.org/10.1080/00207543.2019.1652777.

Mittal, Sameer, Muztoba Ahmad Khan, David Romero, and Thorsten Wuest. 2019. "Smart Manufacturing: Characteristics, Technologies and Enabling Factors." *Proceedings of the Institution of Mechanical Engineers, Part B: Journal of Engineering Manufacture* 233 (5): 1342–61. https://doi.org/10.1177/0954405417736547.

Nylund, Hasse, and Paul H Andersson. 2012. "Digital Manufacturing Supporting Autonomy and Collaboration of Manufacturing Systems." In *Manufacturing System*, edited by Faieza Abdul Aziz. Rijeka: IntechOpen. https://doi.org/10.5772/34596

Pech, Martin, Jaroslav Vrchota, and Jiří Bednář. 2021. "Predictive Maintenance and Intelligent Sensors in Smart Factory: Review." *Sensors* 21 (4): 1470. https://doi.org/10.3390/s21041470

Sony, Michael. 2020. "Pros and Cons of Implementing Industry 4.0 for the Organizations : A Review and Synthesis of Evidence". *Production & Manufacturing Research* 8 (1): 244–72. https://doi.org/10.1080/21693277.2020.1781705.

Sufian, Amr T, Badr M Abdullah, Muhammad Ateeq, Roderick Wah, and David Clements. 2021. "Six-Gear Roadmap towards the Smart Factory." *Applied Sciences* 11 (8): 3568. https://doi.org/10.3390/app11083568

Tao, Fei, Jiangfeng Cheng, Qinglin Qi, Meng Zhang, He Zhang, and Fangyuan Sui. 2018. "Digital Twin-Driven Product Design, Manufacturing and Service with Big Data." *International Journal of Advanced Manufacturing Technology* 94 (9–12): 3563–76. https://doi.org/10.1007/s00170-017-0233-1

2 Indian SMEs – Opportunities and Challenges

Assessing Industry 4.0 Readiness

Virupaxi Bagodi, Deepankar Sinha, and Varsha Bagodi

CONTENTS

2.1 INTRODUCTION

Developing countries such as India have been witnessing the contribution of small and medium enterprises (SMEs) to their economy. SMEs contribute to employment (Ongori 2009; Ramayah et al. 2016) and the balance of trade (Ahmed and Haseen 2017). However, SMEs face intense competition (Kumar et al. 2014; Maurya et al. 2015) and strive hard to produce low cost and good quality products (Bagodi et al. 2020; Ghobadian and Gallear 1996). Competition is from within a country and also from low cost imports (Kumar et al. 2014; Maurya et al. 2015), besides facing the challenges of shortened life cycles and changing technology (Kumar et al. 2014). Factors such as low cost and good quality product factors results in high mortality rates (Setyaningsih 2012).

DOI: 10.1201/9781003200857-2

Many authors emphasise the importance of measuring organisational performance and effectiveness (Singh and Gupta 2016; Bartuševičienė and Šakalytė 2013; Yamin et al. 1999). Initially, financial performance was the primary focus in such measurements (Hofer 1983; Venkatraman and Ramanujam 1986). However, later authors (Velimirović et al. 2011; Cumby and Conrod 2001; Hax and Majluf 1984) suggested including non-financial measurements, and Olaru et al. (2014) proposed four performance measures: finance, processes, human resources, and customers. Bagodi et al. (2020) opined that an overall quality improvement leads to efficient processes, increased financial returns, and better customer and employee relations. However, SMEs lack the ability to identify appropriate performance measures and hence fail to take appropriate action.

Studies (Tarutė and Gatautis 2014; Eniola and Entebang 2015; Klovienė and Speziale 2015) suggest different ways for SMEs to achieve competitiveness including an analysis of their strengths and weaknesses (Deros et al. 2006). However, these studies do not mention state-of-the-art technology and frameworks such as Industry 4.0 implementation.

This chapter demonstrates that Industry 4.0 is crucial for SMEs' growth, but an evaluation of the present state of an SME is required to justify its significance. In this study, a questionnaire survey of 200 SMEs in India was carried out to identify current issues and challenges under four dimensions: process (technology), employee, customer, and financial returns. It aims to address the following:

1. Is the current technology producing the desired results?
2. Are customers satisfied with SMEs' offerings?
3. Are employees motivated?
4. What is the extent of the financial returns or earnings of SMEs?

This chapter is composed of seven sections. The next section discusses the challenges faced by SMEs, and Section 2.3 examines the concept, meaning, and success stories of Industry 4.0 including technology as a driver of customer and employee satisfaction. Sections 2.4 and 2.5 describe the research instrument and analyse the responses collected from 200 respondents, respectively. The results are discussed and their implications in real-world managerial settings are presented in Section 2.6.

2.2 CHALLENGES FACED BY SMES

In this modern-day world, SMEs across the globe are facing many challenges in the form of competition from other producers (Muhammad et al. 2010; Tülüce and Doğan 2014); intense global competition (Khalique et al. 2011; Kumlu 2014); and limited capability to meet the challenges posed by globalisation and liberalisation (Reijonen et al. 2015). SMEs have other limitations such as low productivity and low quality output (Kurniawati and Yuliando 2015; O'Neill et al. 2016), and limited knowledge acquisition and technology management (Bojica and Fuentes 2012; Davenport 2005; Macpherson and Holt 2007). Limited access to capital and finance (Lee et al. 2015; Huda 2012; Cull et al. 2006) and lack of skills for the new environment (Ezzahra et al. 2014; Taylor et al. 2004; Singh et al. 2010b) also dominate the functioning of SMEs. There are resource constraints (Rakićević et al. 2016; Hessels and Parker 2013; Singh et al. 2008a); little government support for sustainable growth (Singh et al. 2010a); a general shortage of information (Suh and Kim 2014; Lee et al. 2012); and the high cost of infrastructure (Laufs and Schwens 2014).

Thus, different factors impact different stages in the organisational life cycle. Different firms have different challenges, primarily financial, technological, skills, and competitiveness. These aspects, along with essential indicators mined from the literature, formed the basis for the questionnaire design.

2.3 INDUSTRY 4.0

The industrial world has witnessed three industrial revolutions and now a fourth termed Industry 4.0. This fourth revolution allows for the smart functioning of an organisation that is effective as well as efficient in producing customised products at less cost (Vaidya et al. 2018). It aims for higher productivity and performance through horizontal and vertical integration (Dalenogare et al. 2018). Resources such as humans, machinery, and other facilities are vertically connected and organisations are horizontally connected in the product value chain through cyber-physical systems (CPS) (Kagermann et al. 2013; Waibel et al.2017). Challenges and issues, as identified by various researchers, during its implementation include: smart decision-making and conciliation mechanisms, high-speed industrial wireless network (IWN) protocols, system modelling and analysis, modularised and flexible artefacts (Wang et al. 2016), manufacturing specific big data and analytics (Thoben et al. 2017), cybersecurity (Rüßmann et al. 2015), and investment (Valdez et al. 2015). Flexibility, cost, efficiency, quality, and competitive advantage are the key benefits to SMEs (Masood and Sonntag 2020).

For the successful implementation of Industry 4.0, people need to acquire additional skills and hence, education and training are vital (Birkel 2019; Erol et al. 2016; Schuh et al. 2018). Unlike computer integrated manufacturing (CIM), it calls for sustained interaction between humans and machines for adding value in production. Therefore, human safety practices such are low worker fatigue, good communication, less risk of accidents etc. (Block and Keller 2015; Kamble et al. 2018; Tupa et al. 2017). A marriage of practices such as lean manufacturing and just-in-time production, and technological developments such as CPS enables value creation in future smart factories (Moktadir et al. 2018; Tortorella and Fettermann 2018). Agile organisations (Hermann et al. 2016) and cultural changes (Davies et al. 2017) are a necessity for the successful implementation of Industry 4.0.

2.4 RESEARCH INSTRUMENT AND SURVEY

2.4.1 Research Instrument

SMEs are highly imperative and are key drivers for the growth of the manufacturing sector (Garengo and Bernardi 2007). Hence, they need to be competitive with good employee morale and highly effective processes (Xenidis and Theocharous 2014). Performance measures are vital for control purposes in both operations and strategic issues (Garengo et al. 2005). In general, financial indicators such as a quick ratio, an inventory turnover ratio, and returns on assets are used to assess organisational performance. Other researchers (e.g. Stojkic and Bosnjak 2019; Sinisammal et al. 2012) carried out studies considering marketing, operations, and human resources as performance indicators. Johnson and Kaplan (1987), Neely et al. (1994), and Ghalayani and Noble (1996) focused on intangible variables (attributes) such as the quality of a product, buyer happiness, and staff morale.

Performance of the processes is the resultant of efficiency in various operations carried out during process. These efficiencies should necessarily result in a reduction in costs and enhanced productivity (Porter 1985). This includes good utilisation of capacity. The enterprises must have the ability to produce good quality products at lower costs and offer excellent customer service (Kaplan 1983; Drury 1990).

SMEs from developed nations have proposed various classes of performance measures, some qualitative and some quantitative: on-time delivery of goods/services, quality goods/services, customer happiness, employee esteem, resource utilisation, and efficient operations (Otley 1997; Maskell 1989; Ittner and Larcker 1998). Singh et al. (2008b) and Kumar et al. (2014) argue that these are relevant even in the Indian context. The process measurement variables include scrap produced during production, quantum of rework, number of defective products, changes in production schedule, and facility deployment (Elbashir et al. 2008); employee morale includes employee attrition, not fulfilling tasks, and being late for work (Abdel-Maksoud 2004); and customer satisfaction includes few complaints, buyer retentiveness, and adding customers (Abdel-Maksoud 2004). Financial soundness is specified using augmented income, an increase in financial gain on assets, and penetration into the market (Dunk 2011).

Thus, the following attributes were found pertinent for the study: scrap generated from the processes/operations; defects produced; quantum of rework; grievances from clientele; alterations in production schedule; staff attrition; absence from or late for work; capacity utilisation; overtime (OT); financial gain and its change rate; cash flow; return on investment (RoI); cost management; bonus to employees; change in sales volume; customer satisfaction, retention, and addition; and market share.

2.4.2 Conduct of Survey

The questionnaire designed for the survey consists of 21 variables placed under four dimensions, namely financial, customer relationship, employee relationship, and processes (reflecting quality). The respondents were asked to assess the items on a 7-point scale. They were also asked to provide additional information about the organisation.

The survey was conducted during January–April 2019. The authors obtained a list of working SMEs from the appropriate authorities such as district industries centres, chambers of commerce, and SME associations. The authors personally delivered the questionnaire to all 262 functioning SMEs, and obtained responses from 200 organisations, a response rate of 75.47%.

2.5 ANALYSIS

Table 2.1 presents the descriptive statistics of the 21 attributes.

2.5.1 CV: Coefficient of Variation

The data statistics show that the CV is below 20% for employee and customer dimensions. However, the CV for the overtime requirement is higher and close to 23%. Discussions with management indicated that overtime is required to meet exigencies, keeping employees happy due to increased income. This finding shows that SMEs have only financial incentives to motivate while peak load assessment and building redundancy (capacity addition) are absent. These aspects speak of lean organisations but may not conform to resilience (Birkie 2016; Chopra and Sodhi 2014; Christopher and Holweg 2011). The CV for the financial factor is above 0.20, indicating volatility in SMEs' financial conditions, some better and some struggling. The question arises, what impacts finances most in SMEs?

Around 7–8% of firms reported inconsistent financial returns and use of overtime to cope with demand. Around 10% reported high levels of scrap, defects, and reworks. This is illustrated in Table 2.2.

Further, Figure 2.1 shows that 25% of respondents opined that their performance was below 70%. These results imply that SMEs are using manual systems and experience variance in their performance. Only 25% of respondents reported good financial returns. The quality practices of eight selected firms were captured, as shown in Table 2.5.

Series 1–7 represent the scores against the variables. Series 1 refers to a score equal to 1 while Series 7 represents the number of respondents indicating a score of 7. Series 6 shows that around 80–100 respondents said that their firm's operations were in a very good state. Around 40–50 respondents opined that they were excellent. These responses show that all four dimensions – finance, technology/process, customer, and employees – are well managed. Thus, 50% of the responses ranged from poor to moderate performance – a band requiring attention. This fact is well supported by the reality that Indian SMEs have high mortality rates (Setyaningsih 2012), and global Indian SMEs face stiff competition.

Table 2.3 shows the correlation matrix among the variables. The results show a high correlation between customer satisfaction and financial returns in terms of profits and RoI. The employee relation attribute directly affects the quality of production – scrap, rework, and defects. These findings suggest that human dependency and lack of automation are causes of poor output (Table 2.4).

A detailed study was carried out in eight selected organisations with a focus on quality management practices and prevalence. Table 2.5 reveals that the organisations had statistical quality control (SQC) in place to measure the defect rates and

TABLE 2.1
Descriptive Statistics of 21 Attributes

Attributes	Sample	Minimum	Maximum	Mean	Std. Dev.	CV
Scrap from production is less (V1)	200	2	7	5.77	1.19	0.21
Less number of defective products in production (V2)	200	2	7	5.87	1.12	0.19
Minimal rework carried out (V3)	200	1	7	5.87	1.15	0.20
Customers' complaints have been considerably reduced (V4)	200	1	7	5.82	1.21	0.21
Alterations in the production schedules are insignificant (V5)	200	1	7	5.81	1.18	0.20
Attrition rate is very low in the organisation (V6)	200	2	7	5.91	1.05	0.18
Less absenteeism is observed among workers (V7)	200	3	7	5.93	1.05	0.18
Workers are punctual to work (V8)	200	1	7	5.90	1.08	0.18
Capacity utilisation of machinery/ equipment is very high (V9)	200	2	7	5.96	1.02	0.17
No need for overtime in your organisation (V10)	200	1	7	5.8	1.33	0.23
Profits have been steady during the last 60 months (V11)	200	1	7	5.58	1.22	0.22
Cash flow has been consistent during the last 60 months (V12)	200	1	7	5.65	1.17	0.21
RoI (yield on capital) has been steady during the last 60 months (V13)	200	1	7	5.69	1.25	0.22
Proper costing prevails in the organisation (V14)	200	1	7	5.76	1.19	0.21
Good bonus has been paid to the workers during the last 60 months (V15)	200	1	7	5.81	1.11	0.19
Consistent increase in sales has been observed during the last 60 months (V16)	200	1	7	5.75	1.24	0.22
Financial gains have consistently increased during the last 60 months (V17)	200	1	7	5.71	1.28	0.22
Satisfied customer base has steadily increased during the last 60 months (V18)	200	1	7	5.93	1.05	0.18
Customer loyalty has steadily grown during the last 60 months (V19)	200	2	7	6.01	1.01	0.17
New clientele has increased during the last 60 months (V20)	200	1	7	5.93	1.16	0.20
Penetration into the market has grown during the last 60 months (V21)	200	2	7	5.95	1.13	0.19

TABLE 2.2
Scores of 21 Variables

Variables	N	1	2	3	4	5	6	7
		Scores						
Scrap from production is less (V1)	200	0	7	5	12	33	90	53
Less number of defective products in production (V2)	200	0	7	2	10	29	96	56
Minimal rework carried out (V3)	200	1	5	3	10	33	87	61
Customers' complaints have been considerably reduced (V4)	200	1	4	7	12	34	79	63
Alterations in the production schedules are insignificant (V5)	200	2	1	7	10	48	67	65
Attrition rate is very low in the organisation (V6)	200	0	2	5	10	40	78	65
Less absenteeism is observed among workers (V7)	200	0	0	7	12	39	73	69
Workers are punctual to work (V8)	200	1	2	4	12	32	88	61
Capacity utilisation of machinery/equipment is very high (V9)	200	0	2	7	5	30	95	61
No need for overtime in your organisation (V10)	200	3	7	5	9	28	84	64
Profits have been steady during the last 60 months (V11)	200	2	4	9	11	50	83	41
Cash flow has been consistent during the last 60 months (V12)	200	2	3	8	11	44	91	41
RoI (yield on capital) has been steady during the last 60 months (V13)	200	3	5	5	11	37	92	47
Proper costing prevails in the organisation (V14)	200	2	3	7	10	39	85	54
Good bonus has been paid to the workers during the last 60 months (V15)	200	1	4	5	6	43	89	52
Consistent increase in sales has been observed during the last 60 months (V16)	200	1	7	7	4	45	79	57
Financial gains have consistently increased during the last 60 months (V17)	200	2	5	9	11	34	85	54
Satisfied customer base has steadily increased during the last 60 months (V18)	200	1	3	4	4	36	93	59
Customer loyalty has steadily grown during the last 60 months (V19)	200	0	2	6	5	29	92	66
New clientele has increased during the last 60 months (V20)	200	2	5	3	6	26	95	63
Penetration into the market has grown during the last 60 months (V21)	200	0	5	5	10	23	90	67

variations in specifications. Four of the eight organisations did not have International Organisation for Standardisation (ISO) certification. Organisations that claimed that a quality management system (QMS) was in place, had no knowledge to support six sigma, lean, and Kaizen, or continuous improvement to address the quality of outputs.

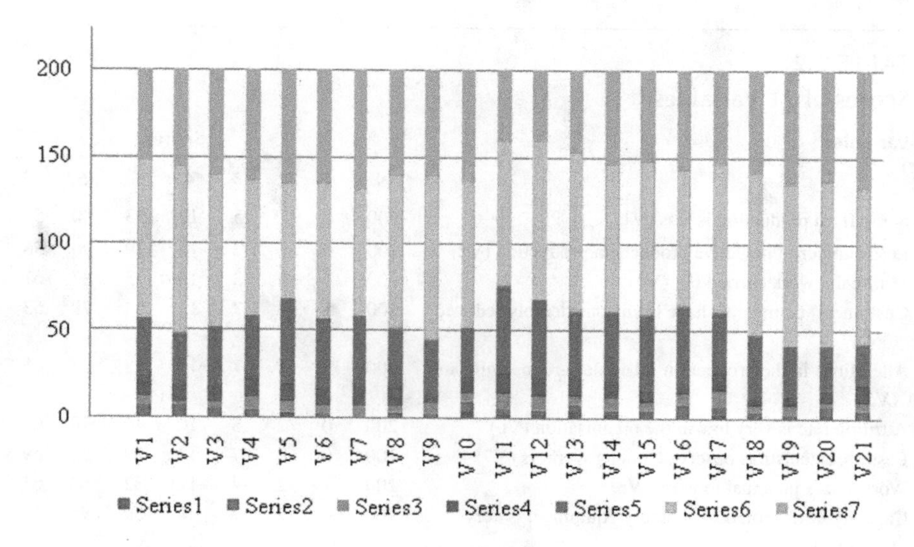

FIGURE 2.1 SME responses.

2.5.2 Technology Index

A technology index (TI) is computed to measure the extent of compliance with customers' requirements. A TI takes the form shown in Equation 2.1:

$$\mathrm{TI} = \prod_{i=1}^{4} T_{Fi} \tag{2.1}$$

where T_{Fi} is the technology success factor. There are four factors – scrap index (SI), defect index (DI), rework index (RI), and complaint index (CI):

$$T_{Fi} = \left(\prod_{j=1}^{7} n_{sj}\right) \wedge (1/i) \tag{2.2}$$

where n refers to the number of respondents assigning a jth score. The scale used in the research instrument ranges from 1 to 7. Equation 2.2 demonstrates that T_{Fi} is the geometric mean of all responses.

Table 2.5 shows the results: TI is lower than the ideal score of 1400, as there are 200 respondents and 7 is the maximum score a respondent can assign. The lowest score is against the generation of scrap meaning the generation of waste, and firms need to be lean (Table 2.6).

Table 2.7 shows that SMEs are operating at 82% efficiency, meaning room for further improvement. The defect ratio is 0.84 and the customer complaint ratio is 0.83, meaning a requirement for proactive measures to reduce defects and hence reduce complaints.

TABLE 2.3

Correlation Matrix among the 21 Variables

Variables	V1	V2	V3	V4	V5	V6	V7	V8	V9	V10	V11	V12	V13	V14	V15	V16	V17	V18	V19	V20	V21
V1	1.00																				
V2	0.77	1.00																			
V3	0.73	0.78	1.00																		
V4	0.71	0.75	0.80	1.00																	
V5	0.67	0.70	0.66	0.80	1.00																
V6	0.51	0.53	0.66	0.65	0.65	1.00															
V7	0.57	0.55	0.58	0.61	0.70	0.79	1.00														
V8	0.50	0.58	0.61	0.69	0.64	0.77	0.81	1.00													
V9	0.60	0.52	0.60	0.63	0.65	0.64	0.77	0.79	1.00												
V10	0.39	0.45	0.45	0.46	0.46	0.55	0.48	0.61	0.54	1.00											
V11	0.54	0.51	0.50	0.55	0.57	0.45	0.47	0.49	0.49	0.46	1.00										
V12	0.51	0.46	0.51	0.54	0.55	0.46	0.48	0.49	0.53	0.46	0.92	1.00									
V13	0.54	0.44	0.51	0.56	0.55	0.43	0.46	0.49	0.55	0.44	0.89	0.90	1.00								
V14	0.46	0.49	0.49	0.55	0.52	0.42	0.42	0.49	0.47	0.44	0.84	0.84	0.86	1.00							
V15	0.45	0.51	0.58	0.56	0.53	0.51	0.44	0.49	0.44	0.39	0.64	0.65	0.67	0.70	1.00						
V16	0.46	0.42	0.44	0.49	0.56	0.36	0.41	0.40	0.42	0.38	0.69	0.69	0.75	0.71	0.72	1.00					
V17	0.43	0.43	0.45	0.46	0.50	0.36	0.38	0.44	0.47	0.45	0.75	0.73	0.79	0.72	0.69	0.84	1.00				
V18	0.36	0.38	0.47	0.45	0.38	0.37	0.29	0.37	0.36	0.25	0.59	0.60	0.61	0.61	0.63	0.60	0.67	1.00			
V19	0.41	0.41	0.45	0.43	0.40	0.32	0.32	0.37	0.39	0.23	0.58	0.59	0.60	0.56	0.67	0.70	0.69	0.79	1.00		
V20	0.34	0.34	0.41	0.40	0.38	0.37	0.26	0.31	0.33	0.30	0.57	0.59	0.63	0.65	0.64	0.66	0.69	0.81	0.72	1.00	
V21	0.40	0.41	0.39	0.43	0.48	0.38	0.40	0.41	0.38	0.41	0.64	0.62	0.63	0.62	0.69	0.71	0.72	0.62	0.72	0.75	1.00

TABLE 2.4
Profile of the Organisations

Organisation	1	2	3	4	5	6	7	8
Type	Small	Small	Medium	Small	Medium	Small	Medium	Small
Workforce	<20	20–50	51–100	20–50	20–50	20–50	101–250	<20
Availability of manager (quality)	√	√	√	√	√	–	√	–
QMS certification	√	–	√	–	√	√	–	–
Application of quality/ continuous improvement techniques	√	√	√	√	√	√	√	√
Quality management system	√	–	–	√	√	√	√	–

TABLE 2.5
Degree of Knowledge and Its Application

Quality improvement strategies and tools	Degree of knowledge					Degree of application				
	NK	PK	TK	GK	EK	NU	RU	MU	FU	EU
ISO9001			4	4		4			4	
Total quality management	8					8				
Six sigma	8					8				
Lean management	8					8				
Kaizen	8					8				

Note: NK: no knowledge; PK: poor knowledge; TK: theoretical knowledge; GK: good knowledge; EK: excellent knowledge; NU: never used; RU: rarely used; MU: moderately used; FU: frequently used; EU: extensively used.

A linear regression analysis with the financial performance score as the dependent variable and process, customer, and employee relationships as the independent variables shows that the latter two variables are significant while the process does not greatly impact finances. The performance process has been found significant at the 10% significance level.

TABLE 2.6
Technology Index (TI) and Its Components

Scrap index (SI)	Defect index (DI)	Rework index (RI)	Compliant index (CI)	Technology index (TI)	Ideal score (IS)
1153	1173	1174	1163	1165	1400

TABLE 2.7
Performance Ratios

Ratio SI/IS	Ratio DI/IS	Ratio RI/IS	Ratio CI/IS	Ratio TI/IS
0.82	0.84	0.84	0.83	0.83

TABLE 2.8
Standardised Total Effect among the Factors

Sl no.	Indicator/factor	Employee indicators	Technology and process indicators	Customer indicators
1	Technology and process indicators	0.8	0	0
2	Customer indicators	0.5	0.5	0
3	Financial indicators	0.59	0.46	0.6

2.5.3 Causality

Structure equation modelling was carried out on the responses obtained to identify causality. AMOS 20 software was used and Table 2.8 shows the effect of one dimension on the other.

The financial performance is influenced almost uniformly by all three dimensions of business – technology and process, customer, and employee. The technology and process dimensions strongly impact on employees and moderately effects on customers. Thus, technology improvement is necessary for the employee dimension to improve as it has a close to 60% impact on finances. Thus, sustainable technology and processes is a crucial enabler to sustainable business.

2.6 RESULTS, DISCUSSIONS, AND MANAGERIAL IMPLICATIONS

Firms are aware of statistical quality control measures but their implementations are reactive in nature, i.e. corrections are made after the detection of errors. This

FIGURE 2.2 Implementation of Industry 4.0.

has resulted in the generation of waste, rework, and complaints. SQC has resulted in around 80% effectiveness, but it is argued that integrating the internet of things (IoT) with the production process can make 100% control possible. The IoT is an enabler of Industry 4.0; it ensures real-time feedback and communication between the system and the operator (Godina and Matias 2018). Ghobakhloo and Fathi (2019) showed how SMEs implemented lean digitised manufacturing systems using concepts of Industry 4.0 and maintained competitiveness.

The SMEs under study had very little knowledge of quality management systems. A QMS enables a framework to be in place for managing and maintaining quality. Six sigma is considered to be one such approach that has a far-reaching impact in ensuring control over errors. The sigma metric of the current process in a firm dictates the type of SQC method to be implemented (Westgard and Westgard 2016). Thus, the discrete application of SQC procedures needs to be replaced with a proper QMS.

A QMS requires the implementation of value stream mapping (VSM) to identify the value adding (VA) and non-value-adding (NVA) activities. Once the process is re-engineered and the correct SQC procedure has been chosen, the wireless sensor network (WSN) and embedded systems (ES) requirements are assessed for the successful implementation of Industry 4.0 and the realisation of its benefits. These stages will lead to the implementation of a digitised lean-agile (DLA) process leveraging information technology. Prerequisites for the implementation of Industry 4.0 in Indian SMEs are illustrated in Figure 2.2.

2.7 CONCLUSIONS

The industrial world has witnessed three revolutions and is now embracing a fourth revolution. This revolution focuses on customisation and lean and agile production through the extensive adoption of the IoT and CPS. The important enabler is the technology adopted. However, Indian SMEs suffer from low adherence between the governing dimensions of their businesses: customers, processes, employees, and finances. This study infers that more than 50% of firms suffer from poor output – defects, scraps, and reworks – and there are inconsistencies in deliverables causing customer complaints. These affect employee motivation and, in turn, customers and finances. Thus, the results show a causality that deviates from conventional beliefs – that is "process and technology impact employees directly which impact the customer satisfaction". This is a shift from the common agreement – that is "process and technology impact customer satisfaction" (Hurley et al. 2014). Both findings are correct but apply to different settings –Indian SMEs are less automated. The chapter introduces a technology index, a metric to assess firms' production effectiveness.

Thus, the adoption of Industry 4.0 by Indian SMEs will lead to more consistency and accuracy as it calls for digitalisation and automation, and does away with the mediating impact of employees. However, the adoption of Industry 4.0 needs a well-defined and optimal process with reduced non-value-added activities, a quality management system followed by automated production stages. The study can be extended to include the identification of constraints and prerequisites to implement sensor and embedded technology-based production processes. This can be the future scope of work.

REFERENCES

Abdel-Maksoud, A. B. 2004. Manufacturing in the UK: Contemporary characteristics and performance indicators. *Journal of Manufacturing Technology and Management* 15 (2): 155–71.

Ahmed, I. and S. Haseen. 2017. Growth of MSME sector and its contribution to exports of India in post reform period. *International Journal of Business Economics & Management Research* 7 (12): 54–65.

Bagodi, V., T. V. Sreenath, and D. Sinha. 2020. A study of performance measures and quality management system in small and medium enterprises in India. *Benchmarking: An International Journal* 28 (4): 1356–89.

Bartuševičienė, I. and E. Šakalytė. 2013. Organizational assessment: Effectiveness vs. Efficiency. *Social Transformations in Contemporary Society* 1: 45–53.

Birkel, H. S., J. W. Veile, J. M. Müller, E. Hartmann, and K.-I. Voigt. 2019. Development of a risk framework for Industry 4.0 in the context of sustainability for established manufacturers. *Sustainability* 11 (2): 384.

Birkie, S. E. 2016. Operational resilience and lean: In search of synergies and trade-offs. *Journal of Manufacturing Technology Management* 27 (2): 185–207.

Block, F. L. and M. R. Keller. 2015. *State of innovation: The US government's role in technology development.* Boca Raton, FL: Taylor & Francis.

Bojica, A. M. and M. D. M. Fuentes. 2012. Knowledge acquisition and corporate entrepreneurship: Insights from Spanish SMEs in the ICT sector. *Journal of World Business* 47: 397–408.

Chopra, S. and M. S. Sodhi. 2014. Reducing the risk of supply chain disruptions. *MIT Sloan Management Review* 3: 73–81.

Christopher, M. and M. Holweg. 2011. Supply Chain 2.0: Managing supply chains in the era of turbulence. *International Journal of Physical Distribution & Logistics Management* 41 (1): 63–82.

Cull, R., L. E. Davis, N. R. Lamoreaux, and J. L. Rosenthal. 2006. Historical financing of small- and medium-size enterprises. *Journal of Banking & Finance* 30: 3017–42.

Cumby, J. and J. Conrod. 2001. Non-financial performance measures in Canadian biotechnology industry. *Journal of Intellectual Capital* 2 (3): 261–72.

Dalenogare, L.S., G. B. Benitez, N. F. Ayala, and A. G. Frank. 2018. The expected contribution of Industry 4.0 technologies for industrial performance. *International Journal of Production Economics* 204: 383–94.

Davenport, S. 2005. Exploring the role of proximity in SME knowledge acquisition. *Research Policy* 34: 683–701.

Davies, R., T. Coole, and A. Smith. 2017. Review of socio-technical considerations to ensure successful implementation of Industry 4.0. *Procedia Manufacturing* 11: 1288–95.

Deros, B. M., S. R. M. Yusof, and A. M. Salleh. 2006. A benchmarking implementation framework for automotive manufacturing SMEs. *Benchmarking: An International Journal* 13 (4): 396–430.

Drury, C. 1990. Cost control and performance measurements in an AMT environment. *Management Accounting*, 68 (10): 40–44.

Dunk, A. S. 2011. Product innovation, budgetary control, and the financial performance of firms. *The British Accounting Review* 43: 102–11.

Elbashir, M. Z., P. A. Collier, and M. J. Davern. 2008. Measuring the effects of business intelligence systems: The relationship between business process and organizational performance. *International Journal of Accounting Information Systems* 9: 135–53.

Eniola, A. A. and H. Entebang. 2015. SME firm performance-financial innovation and challenges. *Procedia – Social and Behavioral Sciences* 195: 334–42.

Erol, S., A. Jäger, P. Hold, K. Ott, and W. Sihn. 2016. Tangible Industry 4.0: A scenario-based approach to learning for the future of production. *Procedia CIRP* 54: 13–18.

Ezzahra, K. F., R. Mohamed, T. Omar, and T. Mohamed. 2014. Training for effective skills in SMEs in Morocco. *Procedia – Social and Behavioral Sciences* 116: 2926–30.

Garengo, P. and G. Bernardi. 2007. Organizational capability in SMEs. *International Journal of Productivity and Performance Management* 56 (5/6): 518–32.

Garengo, P., S. Biazzo, and U. S. Bititci. 2005. Performance measurement systems in SMEs: A review for a research agenda. *International Journal of Management Reviews* 7 (1): 25–47.

Ghalayini, A. M. and J. S. Noble. 1996. The changing basis of performance measurement. *International Journal of Operations and Production Management* 16: 63–80.

Ghobadian, A. and D. N. Gallear. 1996. Total quality management in SMEs. *Omega* 24 (1): 83–106.

Ghobakhloo, M. and M. Fathi. 2019. Corporate survival in Industry 4.0 era: The enabling role of lean-digitized manufacturing. *Journal of Manufacturing Technology Management* 31 (1): 1–30.

Godina, R. and J. C. Matias. 2018. Quality control in the context of Industry 4.0. In *International joint conference on industrial engineering and operations management*, 177–87. Cham: Springer.

Hax, A. C. and N. S. Majluf. 1984. *Strategic management: An integrative perspective.* Englewood Cliffs, NJ: Prentice-Hall.

Hermann, M., T. Pentek, and B. Otto. 2016. Design principles for industries 4.0 scenarios. In *2016 49th Hawaii International Conference on System Sciences (HICSS)*, pp. 3928–3937. IEEE, 2016.

Hessels, J. and S. C. Parker. 2013. Constraints, internationalization and growth: A cross-country analysis of European SMEs. *Journal of World Business* 48: 137–48.

Hofer, C. W. 1983. ROVA: A new measure for assessing organizational performance. In *Advances in strategic management*, ed. R. Lamb, 43–55. New York: Jai Press.

Huda, A. N. 2012. The development of Islamic financing scheme for SMEs in a developing country: The Indonesian case. *Procedia – Social and Behavioral Sciences* 52: 179–86.

Hurley, R., M. T. Jan, and K. Abdullah. 2014. The impact of technology CSFs on customer satisfaction and the role of trust: An empirical study of the banks in Malaysia. *International Journal of Bank Marketing* 32 (5): 429–47.

Ittner, C. D. and D. F. Larcker. 1998. Innovations in performance measurement: Trends and research implications. *Journal of Management Accounting Research* 10: 205–37.

Johnson, H. T. and R. S. Kaplan. 1987. The rise and fall of management accounting [2]. *Strategic Finance* 68 (7): 22.

Kagermann, H., W. Wahlster, and J. Helbig. 2013. Recommendations for implementing the strategic initiative industrie 4.0: Final report of the industrie 4.0 working group. doi:10.13140/RG.2.1.1205.8966 (accessed 3 Nov 2020).

Kamble, S. S., A. Gunasekaran, and S. A. Gawankar. 2018. Sustainable Industry 4.0 framework: A systematic literature review identifying the current trends and future perspectives. *Process Safety and Environmental Protection* 117: 408–25.

Kaplan, R. S. 1983. Measuring performance: A new challenge for managerial accounting research. *The Accounting Review* 18 (4): 686–705.

Khalique, M., A. H. M. Isa, J. A. N. Shaari, and A. Ageel. 2011. Challenges faced by the small and medium enterprises (SMEs) in Malaysia: An intellectual capital perspective. *International Journal of Current Research* 3 (6): 398–401.

Kloviené, L. and M-T. Speziale. 2015. Is performance measurement system going towards sustainability in SMEs. *Procedia – Social and Behavioral Sciences* 213: 328–33.

Kumar, R., R. K. Singh, and R. Shankar. 2014. Strategy development by Indian SMEs for improving coordination in supply chain: An empirical study. *Competitiveness Review* 24 (5): 414–32.

Kumlu, Ö. 2014. The effect of intangible resources and competitive strategies on the export performance of small and medium sized enterprises. *Procedia – Social and Behavioral Sciences* 150: 24–34.

Kurniawati, D. and H. Yuliando. 2015. Productivity improvement of small scale medium enterprises (SMEs) on food products: Case at Yogyakarta Province, Indonesia. *Agriculture and Agricultural Science Procedia* 3: 189–94.

Laufs, K. and C. Schwens. 2014. Foreign market entry mode choice of small and medium-sized enterprises: A systematic review and future research agenda. *International Business Review* 23: 1109–126.

Lee, N., H. Sameen, and M. Cowling. 2015. Access to finance for innovative SMEs since the financial crisis. *Research Policy* 44: 370–80.

Lee, Y., J. Shin, and Y. Park. 2012. The changing pattern of SME's innovativeness through business model globalization. *Technological Forecasting & Social Change* 79: 832–42.

Macpherson, A. and R. Holt. 2007. Knowledge, learning and small firm growth: A systematic review of the evidence. *Research Policy* 36: 172–92.

Maskell, B. 1989. Performance measurement for world class manufacturing-3. *Management Accounting*: 48–50.

Masood, T. and P. Sonntag. 2020. Industry 4.0: Adoption challenges and benefits for SMEs. *Computers in Industry* 121: 1–12. doi:10.1016/j.compind.2020.103261.

Maurya, U. K., P. Mishra, S. Anand, and N. Kumar. 2015. Corporate identity, customer orientation and performance of SMEs: Exploring the linkages. *IIMB Management Review* 27: 159–74.

Moktadir, M. A., S. M. Ali, S. Kusi-Sarpong, and M. A. A. Shaikh. 2018. Assessing challenges for implementing Industry 4.0: Implications for process safety and environmental protection. *Process Safety and Environmental Protection* 117: 730–41.

Muhammad, M. Z., A. K. Char, M. R. Yasoa, and Z. Hassan. 2010. Small and medium enterprises (SMEs) competing in the global business environment: A case of Malaysia. *International Business Research* 3 (1): 66–75.

Neely, A., F. Mills, K. Platts, and M. Gregory. 1994. Realising strategy through measurement. *International Journal of Operations and Production Management* 14: 140–52.

O'Neill, P., A. Sohal, and C. W. Teng. 2016. Quality management approaches and their impact on firms' financial performance: An Australian study. *International Journal of Production Economics* 171: 381–93.

Olaru, M., I. C. Pirnea, A. Hohan, and M. Maftei. 2014. Performance indicators used by SMEs in Romania, related to integrated management systems. *Procedia – Social and Behavioral Sciences* 109: 949–53.

Ongori, H. 2009. Role of information communication technologies adoption in SMEs: Evidence from Botswana. *Research Journal of Information Technologies* 1 (2): 79–85.

Otley, D. 1997. Better performance management. *Management Accounting*, 75 (1): 44.

Porter, M. E. 1985. *Competitive advantage*. New York: Free Press.

Rakićević, Z., J. Omerbegović-Bijelović, and D. Lečić-Cvetkovic. 2016. A model for effective planning of SME support services. *Evaluation and Program Planning* 54: 30–40.

Ramayah, T., N. S. Ling, S. K. Taghizadeh, and S. A. Rahman. 2016. Factors influencing SMEs website continuance intention in Malaysia. *Telematics and Informatics* 33: 150–64.

Reijonen, H., S. Hirvonen, G. Nagy, T. Laukkanen, and M. Gabrielsson. 2015. The impact of entrepreneurial orientation on B2B branding and business growth in emerging markets. *Industrial Marketing Management* 51: 35–46.

Rüßmann, M., M. Lorenz, and P. Gerbert et al. 2015. Industry 4.0: The future of productivity and growth in manufacturing industries. https://inovasyon.org/images/Haberler/bcg-perspectives_Industry40_2015.pdf (accessed 9 April 2019).

Schuh, G., C. Dölle, C. Mattern, M.-C. Modler, and S. Schloesser. 2018. Studying technologies of Industry 4.0 with influence on product development using factor analysis. In *Proceedings of NordDesign*, Linköping, Sweden, 14–17 August, 2018. RWTH Aachen University.

Setyaningsih, S. 2012. Using cluster analysis study to examine the successful performance entrepreneur in Indonesia. *Procedia Economics and Finance* 4: 286–98.

Singh, R. K., S. K. Garg, and S. G. Deshmukh. 2008a. Strategy development by SMEs for competitiveness: A review. *Benchmarking: An International Journal* 15 (5): 525–47.

Singh, R. K., S. K. Garg, and S. G. Deshmukh. 2008b. Competency and performance analysis of Indian SMEs and large organizations: An exploratory study. *Competitiveness Review: An International Business Journal* 18 (4): 308–21.

Singh, R. K., S. K. Garg, and S. G. Deshmukh. 2010a. Strategy development by small scale industries in India. *Industrial Management & Data Systems* 110 (7): 1073–93.

Singh, R. K., S. K. Garg, and S. G. Deshmukh. 2010b. The competitiveness of SMEs in a globalized economy: Observations from China and India. *Management Research Review* 33 (1): 54–65.

Singh, S. and V. Gupta. 2016. Organizational performance research in India: A review and future research agenda. In *The Sixth Indian Council for Social Science Research (ICSSR) survey of psychology in India*, ed. Misra, G. New Delhi: Oxford Publishing.

Sinisammal, J., P. Belt, J. Harkonen, M. Mottonen, and S. Vayrynen. 2012. Successful performance measurement in SMEs through personnel participation. *American Journal of Industrial and Business Management* 2: 30–8.

Stojkic, Z. and I. Bosnjak. 2019. An overview of performance measurement methods in SMEs. In *Proceedings of the 30th DAAAM International Symposium*, 518–24. Vienna, Austria. doi:10.2507/30th.daaam.proceedings.070.

Suh, Y. and M. S. Kim. 2014. Internationally leading SMEs vs. internationalized SMEs: Evidence of success factors from South Korea. *International Business Review* 23: 115–29.

Tarutė, A. and R. Gatautis. 2014. ICT impact on SMEs performance. *Procedia – Social and Behavioral Sciences* 110: 1218–25.

Taylor, M. J., J. Mcwilliam, D. England, and J. Akomode. 2004. Skills required in developing electronic commerce for small and medium enterprises: Case based generalization approach. *Electronic Commerce Research and Applications* 3: 253–65.

Thoben, K. D., S. Wiesner, and T. Wuest. 2017. Industrie 4.0 and smart manufacturing: A review of research issues and application examples. *International Journal of Automation and Technology* 11 (1): 4–16.

Tortorella, G. L. and D. Fettermann. 2018. Implementation of Industry 4.0 and lean production in Brazilian manufacturing companies. *International Journal of Production Research* 56 (8): 2975–87.

Tülüce, N. S. and İ. Doğan. 2014. The impact of foreign direct investments on SMEs' development. *Procedia – Social and Behavioral Sciences* 150: 107–15.

Tupa, J., J. Simota, and F. Steiner. 2017. Aspects of risk management implementation for Industry 4.0. *Procedia Manufacturing* 11: 1223–30.

Vaidya, S., P. Ambad, and S. Bhosle. 2018. Industry 4.0: A glimpse. *Procedia Manufacturing* 20: 233–38.

Valdez, A. C., P. Brauner, A. K. Schaar, A. Holzinger, and M. Ziefle. 2015. Reducing complexity with simplicity: Usability methods for industry 4.0. In Proceedings of 19th Triennial Congress of the IEA, Melbourne 9–14 August.

Velimirović, D., M. Velimirović, and R. Stankovića. 2011. Role and importance of key performance indicators measurement. *Serbian Journal of Management* 6 (1): 63–72.

Venkatraman, N. and V. Ramanujam. 1986. Measurement of business performance in strategy research: A comparison of approaches. *The Academy of Management Review* 11 (4): 801–14.

Waibel, M. W., L. P. Steenkamp, N. Moloko, and G. A. Oosthuizen. 2017. Investigating the effects of smart production systems on sustainability elements. *Procedia Manufacturing* 8: 731–37.

Wang, S., J. Wan, D. Li, and C. Zhang. 2016. Implementing smart factory of industrie 4.0: An outlook. *International Journal of Distributed Sensor Networks* 3159805: 1–10. doi:10.1155/2016/3159805.

Westgard, J. O. and S. A. Westgard. 2016. Six sigma quality management system and design of risk-based statistical quality control. *Clinics in Laboratory Medicine* 37 (1): 85–96.

Xenidis, Y. and K. Theocharous. 2014. Organizational health: Definition and assessment. *Procedia Engineering* 85: 562–70.

Yamin, S., A. Gunasekaran, and F. T. Mavondo. 1999. Relationship between generic strategies, competitive advantage and organizational performance: An empirical analysis. *Technovation* 19: 507–18.

3 Paradigm Shift in Construction Processes with Industry 4.0

Sayali Sandbhor, Kirti Ruikar, and Shalaka Hire

CONTENTS

3.1 INTRODUCTION

The world is witnessing the growth of digitisation at an accelerated pace. The ability of digitisation to improve organisational efficiency and effectiveness, and create new opportunities has propelled it to prominence. Digitalisation has been recognised as a new approach to retaining competitive advantage and competitiveness in industries such as banking, manufacturing, and retailing (Osunsanmi, Aigbavboa, and Oke 2018). Although the use of this obviously beneficial technology has been embraced by these other industries, the construction industry has yet to fully embrace it in providing its services. The complex nature of the construction industry requires more advanced and efficient methods that will help to improve productivity. Even so, the industry has already started to use advanced techniques for processes such as designing, cost management, time management, and quality management. These techniques mainly encompass artificial intelligence (AI), robotics, cloud computing, virtual reality (VR), and sensor technologies.

DOI: 10.1201/9781003200857-3

Massive and complex construction processes generate large volumes of data and information. It has also been stated that a construction professional's success is linked to the quality of the information available to them. Having vast information on any project is of no use unless it is shared with the right people at the right time. The fourth Industrial Revolution, i.e. Industry 4.0, is based on the digital revolution, which connects people and technology (Alaloul et al. 2020). Industry 4.0 refers to the rise in manufacturing sector digitisation aimed at facilitating contact between commodities, firms, and users. The industrial sector in Germany gave birth to Industry 4.0 (Yang Lu 2017). The processes involved in the manufacturing sector are mostly repetitive, whereas in the construction sector each project is unique in nature, involving non-repetitive work. Construction Industry 4.0 is focused on the concepts of building smart construction sites, simulation, and virtual data storage, which help construction firms to organise and compare data from different stages of a project as well as data from end users after a project has been completed, in order to deliver quality outputs that provide a faster, more flexible, and lower-cost construction project (Oesterreich and Teuteberg 2016). Despite the advantages of Industry 4.0, its adoption has been slow in the construction sector. Many studies are being conducted to identify the challenges to adopting Industry 4.0 (Alaloul et al. 2020). Industry is aware of the significance of the digitisation megatrend; however, the issue is with implementation (Schober and Hoff 2016). Additionally, virtual data storage is not without its own challenges. Voluminous data sets are largely unstructured, requiring intelligent input into classifying and contextualising the data so that meaningful inferences can be drawn. Also, the move to digitisation means that there is no dearth in the quantity of data. So, the challenges are now associated with sifting through the data, so that the noise can be cut out to maintain the quality of the data. This practice would present numerous opportunities in the future. The benefits of "data" management are spread across life cycle processes and are applicable to various objectives, such as managing energy usage, site safety, improving decision-making, reducing costs, and maintaining quality throughout the life cycle.

The focus of this chapter is on the opportunities and challenges of managing data in the areas of safety management in a construction project. The rapid increase in the number of on-site construction accidents indicates that safety requires special attention in terms of digitisation. Safety management on a typical construction site involves systematic processes such as hazard identification and solution, safety training, inspection, and monitoring. All these processes generate vast amounts of data on different types of hazards, the causes of hazards, their scenarios along with their mitigation, standard safety rules and procedures, monitoring and inspection records, as well as information on the people engaged in the overall process. For the effective functioning of safety planning processes, the data generated needs to be curated and maintained, to drive quality outcomes. It is evident that site safety data management practices are *ad hoc* at best. For instance, current safety practices are text based, check sheet type, and manual. All these practices lead to disconnected data sets, so the challenge is not simply limited to structuring the data, but it is also inherently associated with the culture of creating "disjointed pockets of *ad hoc* multi-media-based data sets" that lead to forensic gaps. At its worst, the consequence of this practice leads to loss of life, which is an expensive price to pay. Additionally, poor safety practices lead to accidents, cost lives, result in delays, and consequently affect productivity. Gaps in the data maintenance regime (including paper and digital records, file types, formats, storage, archives) make a forensic "resolution" a challenge. Given that current safety methods are inefficient, there is an urgent need to adopt more effective practices that promote new ways of examining existing data to drive better decisions and promote best practices. A typical site management process consists of connected interactions between cyber-physical systems. This system is composed of digital, analogue, physical, and interaction between people who collaborate to perform a specific task (e.g. site safety function) through integrated physics and logic. In line with this, the study proposes a framework for intelligent safety management, which would be addressed through the collection, classification, and codification of site safety data, so that "knowledgeable" inferences are drawn.

3.1.1 Rationale

The construction sector is widely regarded as having one of the worst records in terms of workplace health and safety. Construction is one of the most challenging sectors, with around half of all injuries and worker fatalities occurring in the workplace (Ganah and John 2015). Poor safety practices on construction sites usually result in injuries or fatalities. Despite the fact that the construction industry in India is booming, the government has taken no steps to ensure that safety rules and regulations are followed (Kanchana, Sivaprakash, and Joseph 2015)

Construction workers in India account for 7.5% of the global workforce, and 16.4% of all fatal workplace accidents occur on construction sites. According to the Occupational Safety and Health Administration (OSHA), 1 out of every 10 construction workers is injured each year. Falling hazards are the primary cause of injury on construction sites, according to the OSHA. Based on the OSHA's fatal four hazards (OSHA 2011), Figure 3.1 illustrates the OSHA fatal four along with percentage and cause.

According to an article published in the *Indian Express* in 2019, over 48,000 workers in India die each year as a result of work-related accidents, with 38 of these occurring every day in the construction industry. Every year, over 150,000 people are injured on construction projects, based on the current Bureau of Labor Statistics. Between 2008 and 2012, the minimum number of individuals killed in India's construction industry was 11,614 a year (Patel and Jha 2016). Labour safety is vital because it promotes a positive and productive work environment. Management, on the other hand, pays the least attention to safety (Hire, Sandbhor, and Ruikar 2021b). Studies conducted on the causes of accidents include lack of personal protective equipment, lack of knowledge and training on equipment, poor safety management, and management ignorance of safety provisions. Statistics and data sets indicate that despite conventional safety practices, the number of accidents is increasing.

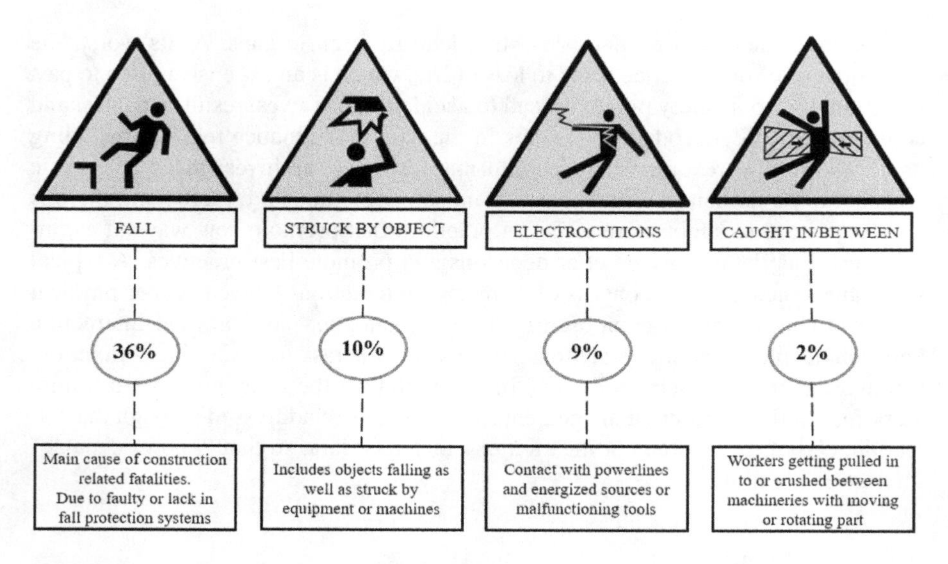

FIGURE 3.1 Fatal four by OSHA (OSHA 2011).

3.1.2 State-of-the-Art Industry 4.0 in Construction Safety

Industry 4.0 promises increased efficiency by combining digital manufacturing systems with the analysis and collaboration of all data produced in a smart world. The trends and developments of the digital systems in Industry 4.0, especially for construction safety management, are extracted from the Scopus database using a set of four keywords including "Industry 4.0" or "Industrial Revolution" and "Construction" or "Construction 4.0" and "Safety". The search uncovered 71 documents, of which only 46 are on engineering and were published from 2010 to 2021. The first publication in the domain was in 2015, with the number of publications in the domain of Industry 4.0 in construction safety management seeing a sudden increase in 2020. The United States and Italy are leading in research with five publications to their credit. Canada, China, and Germany have four publications each. Countries including Austria, Malaysia, South Africa, Switzerland, the United Arab Emirates, and the United Kingdom are improving their contribution through research, having two publications each to their credit. With only one publication in the proposed domain, Indian researchers have large scope in this field and can lead the development of innovative and effective safety approaches (Scopus data accessed 3 May 2021).

3.2 REPRESENTATIVES OF INDUSTRY 4.0 IN CONSTRUCTION SAFETY

Keywords reflect the key elements of the research. Critically observing the keywords selection, various crucial tools, techniques, and processes can be identified. Figure 3.1 shows the network of different keywords used in the publications derived from a systematic search conducted on the selected database. With a threshold of

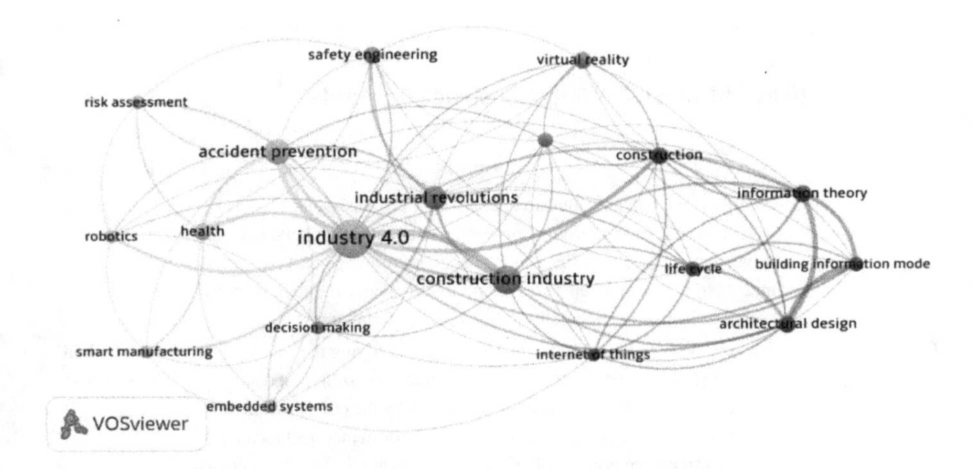

FIGURE 3.2 Keyword network prepared using the VOSviewer tool (SCOPUS data accessed on 3 May 2021).

four, 13 of the most frequently used keywords were obtained and are presented in Figure 3.2 with their interrelations. These sets of keywords, which represent various digitised processes, tools, and technologies such as building information modelling (BIM), robotics, virtual reality, internet of things (IOT), and embedded systems, are the representatives and torchbearers for adopting Industry 4.0 in construction safety management–related practices.

The observations of the keyword network are the driving force for the need to systematically develop an understanding of technology adoption and usage in construction safety management. AI, data mining, radio-frequency identification (RFID), cellular networks, VR, automation and robotics, sensor technology, global positioning system (GPS), geographic information system (GIS), augmented reality, wireless, and simulation technology are some of the digital innovations used in safety planning (Z. Zhou, Irizarry, and Li 2013). Table 3.1 shows some of the advanced techniques and processes proposed and used in construction safety planning–related research. Each of these representatives involves the utilisation of various tools and techniques for performing safety checking and mitigating processes. Figure 3.3 illustrates some of the available platforms utilised for digitised safety planning and their function in safety planning. Along with these tools, coding platforms can also help in the automatic safety checking of construction sites. An overview of the software applications developed by different countries and government authorities for automated compliance with each country's building code is described in the study by Narayanswamy, Liu, and Al-Hussein (2019).

The representatives outlined in Table 3.1 show that, in the last two decades, a variety of digital safety processes, techniques, and tools have been studied and can be utilised to improve safety standards. Technology exists to bridge the gap between on-site experiences, educate employees about best safety practices, and compile health and safety data into real-time analytics to pinpoint how to

TABLE 3.1
Representatives of Industry 4.0 in Construction Safety

Digital representative	Description	Reference
AI	AI helps to predict and enhance operator fitness by assisting decision-making and recognising cognitive tasks. Smart personal protective equipment (PPE) can connect with motion smart containers and smart cranes in intelligent container ports to deliver alerts in risky scenarios such as accidentally walking under a container.	Romero et al. (2018)
Cloud computing	Data collected can be useful for improving and verifying the data's reliability. Giving the wristband to a co-worker, adjusting past data regarding alerts and environmental conditions by the company, calculating incorrect performance, and collecting data from incorrect locations are all examples of issues with incorrect computing. External inspectors may use the cloud to conduct inspections and ensure enforcement. Furthermore, cloud computing may be used to schedule training.	Barata and da Cunha (2019)
VR	VR is a range of hardware and software technologies that are used to create interactive, real-time, three-dimensional computer applications. For example, the Building Management Simulation Centre is used to train construction personnel in a safe and realistic virtual construction site. The tool has four functions: collision detection, terrain following, geometry selection, and 3D tape measurement. Job safety assessments are supported by realistic models of current work conditions. For better training, the device includes virtual images, animation, and a 3D interactive environment.	Zhou, Irizarry, and Li (2013)
Ultra-wide-band (UWB)	UWB helps to develop a real-time building safety management system to track workers at the construction site. This system enabled the building of a virtual wall in the region of a construction project's fall hazard and danger area. An alert emerged during this scene as workers were trespassing or staying too close to the fence. The signal was transferred to another UWB receiver sensor from a sender sensor. This challenge has the drawback of not producing a solid signal capable of travelling through more than two thick barriers. A single UWB receiver is unable to effectively cover a big area.	Giretti et al. (2009)
BIM	BIM is often used on construction sites because of its ability to simulate technical details about a building's cost estimate, material inventories, and completion time. As a result, BIM can be considered one of the first inventions that set the stage for building. 4.0 out of 5 stars BIM and protection rules are used to design automated identification of dangerous design factors in buildings. By integrating BIM with design safety regulations, this research aims to develop an approach to design for safety, which can identify safety issues arising from a design. BIM describes the identification and avoidance of automated scaffolding-based safety hazards by incorporating work orders and structures into safety. Construction models and schedules created with BIM are automatically checked for protection. For safety checks, BIM has been combined with other technologies such as Solibri Model Checker (SMC) and GPS.	Osunsanmi, Aigbavboa, and Oke (2018) Hongling et al. (2016) Kim et al. (2016) Zhang et al. (2013) Arslan et al. (2019) Hossain and Ahmed (2019)

(*Continued*)

TABLE 3.1 (CONTINUED)
Representatives of Industry 4.0 in Construction Safety

Digital representative	Description	Reference
Sensing and warning technologies	Positioning sensors and routers to capture workers' data in real-time were developed. The developed software analysed the data. A warning was given to the machine if the worker remained close to the predefined fall danger zone. A wireless monitoring technology and a local server operating a data collection software network, a real-time website, introduced the programme. Also, mobile devices can be used for monitoring environment sensors, object's sensors, and biometric sensors.	Naticchia, Vaccarini, and Carbonari (2013) Barata and da Cunha (2019)
GPS and GIS	The safety monitoring and control system was created to enable the safety manager to monitor and control workers on the job site from a distance. Each worker was given a GPS tracking system with a specific device name and code. To track and control workers in the event of a fall threat, a control panel is required. However, this scheme is costly, and it necessitates a constant wi-fi. GIS stores comprehensive information about the area, including weather, topography, thermal comfort, and access route planning. With the surrounding topography and schedule, a 3D model can be created.	Zhou et al. (2012)
RFID	It is the use for radio-frequency through utilization and the gathering of data placed on a label to an object. IOT, RFID, cyber-physical networks, and other technologies are used in the smart factory. In a working environment, the RFID device can be used to detect collisions. Warning systems can help avoid collisions and falls. RFID systems may be used to prevent unauthorised entry to a site.	Osunsanmi, Aigbavboa, and Oke (2018) Barata and da Cunha (2019)
Encoding and programming languages	The core OWL (ontology language) is used to build a model for information encoding. This software extracts provisions for constructing safety regulations that are written in Semantic Web Rule language. Numerous encoding and programming languages are used in this model. This model employs a variety of encodings and programming languages. This method is very complicated and difficult for construction workers to comprehend.	Ying Lu et al. (2015)
Cyber-physical system	Based on the cyber-physical system, a safety management is proposed. The device creates risk data synchronisation mapping between virtual and physical construction sites using scene reconstruction design, data understanding, data communication, and data processing modules. On-site risks will be established and handled, including personnel, mechanical, and other threats.	Jiang, Ding, and Zhou (2020)

mitigate or eliminate risks completely. In this digital age, data generation, data storage, data separation, and data use to improve construction safety are critical. This chapter presents the conventional safety planning practices and provides a theoretical framework using Industry 4.0 principles to overcome the drawbacks of these practices.

FIGURE 3.3 Available platforms and their safety functions for digitised safety planning.

3.3 CONVENTIONAL SAFETY PRACTICES IN CONSTRUCTION

Safety planning is crucial in construction project management because it helps to avoid unnecessary costs and delays caused by unintended accidents (Chantawit et al. 2005). To transfer the conventional safety practices into advanced digitised procedures, it is imperative to understand the existing safety practices and identify loopholes.

3.3.1 General Safety Management Framework

Safety planning ensures that safety, along with costs, schedules, efficiency, and other critical job objectives, is taken into account. The identification of all safety

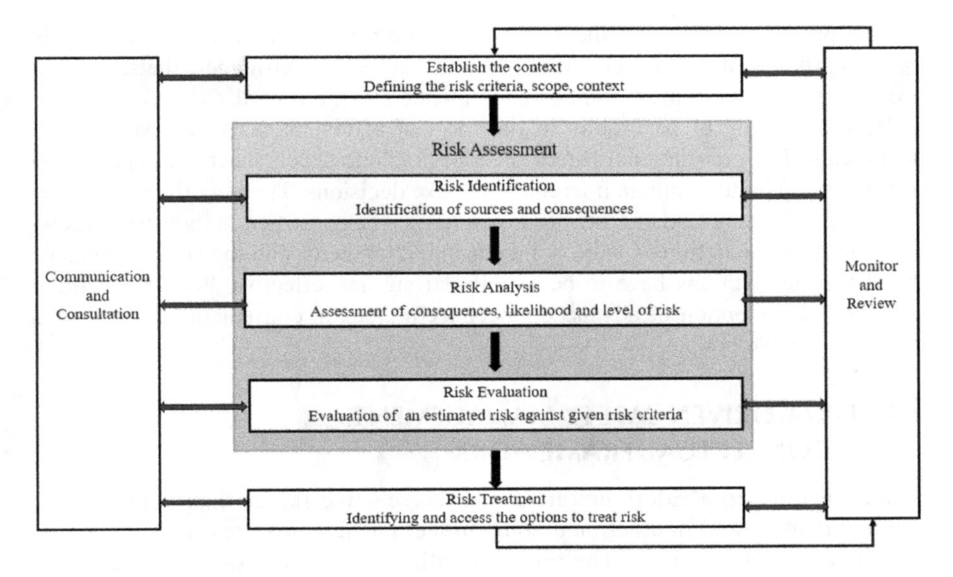

FIGURE 3.4 Conventional safety management process flow.

hazards and an evaluation of the risk associated with each hazard are key parts of safety management. Procedures for risk management involve continuous monitoring with routine evaluations of safety results, and continuous improvement of the safety management system's effectiveness. Identifying risk, precautionary and remedial measures, safety training, and safety monitoring are the key parameters of any safety management process. There are no definitive guidelines for assessing, communicating, and controlling workplace health and safety risks. However, there are some general principles that should be followed (Yulistianingsih et al. 2012). Regardless of variations in methods or sectors, the majority of risk management methodologies follow the same basic concepts. On the basis of ISO 31000, the key process of conventional site safety management is given in Figure 3.4, which includes establishing the context, hazard identification, risk analysis, risk evaluation, and risk treatment. All the components of a given framework take place in the physical environment.

The intricacy of risk evaluation, risk communication, and risk management is largely determined by factors such as the organisation's size, workplace situations, and the scope, complexity, and effect of the risks to which it is exposed. Hazard must, however, be measured at any point in a construction project's life cycle. Designers in health and safety risk evaluation activities should be utilised to design out elements of a building or structure that pose a threat to the health and safety of construction workers. Traditionally, safety managers have supervised construction sites based on their previous work experience and visual observation. Collecting reports on a construction project's progress can be costly and time-consuming. The contractor's engineers on site collect the necessary on-site progress data, and the site works' progress is recorded by manually updating the construction schedule, which is backed up by images, the site diary, progress meeting minutes, and correspondence (Vacanas et al. 2015).

Record-keeping of hazards is in the form of safety manuals and paper-based documents. The conventional safety practices include check sheets or paper-based processes and require human inferences to make decisions. These methods are time-consuming and may lead to error-prone results. As the construction industry is growing rapidly, the quantum of work is increasing, thus increasing the risk. To manage the risks, the solutions have to be equally strong and effective. To achieve these solutions, the adoption of advanced safety techniques in construction processes is the most suitable option.

3.4 DATA-DRIVEN SAFETY MANAGEMENT: A CONCEPTUAL FRAMEWORK

Crucial actions in a safety culture are often reactive rather than constructive. Communication and interaction among project participants are critical to a construction project's success. The problem with most existing safety initiatives is that, while the data they collect is useful for looking back on previous success, it lacks the information needed to make concrete decisions about future best practices. The majority of safety systems depend on a one-way feedback loop that evaluates accidents after they occur. In the building and related sectors, this reactive approach to safety management tends to place workers at risk of injury or even death. Predicting accidents and taking appropriate action to minimise risks before they occur is the best way to avoid them. The proposed framework is designed for safety management throughout the project life cycle. It is understood that the practice of safety management takes place within cyber-physical systems. Hence, the interaction of a cyber-physical element in the safety framework is later explained in Section 3.4.2.

3.4.1 Proposed Framework for Proactive Safety Management

Advanced techniques have the ability to identify and manage hazards prior to construction. Industry 4.0 is about bringing customised solutions to persistent problems that need data-driven and information and communications technology (ICT)-based approaches. Taking inspiration, a conceptual framework is developed for a proactive safety checking system that can be run virtually for construction operations (Figure 3.5), understanding from which can be implemented again in the virtual scenario to improve the safety of the model. The framework is developed based on a robust foundation of prior studies conducted (Hire et al. 2021a,b). This improved safety plan is implemented in a real-time construction operation. As the model is based on concepts of Industry 4.0 in construction, it focuses on digitised safety planning with improved communication and coordination among the different actors involved in the process.

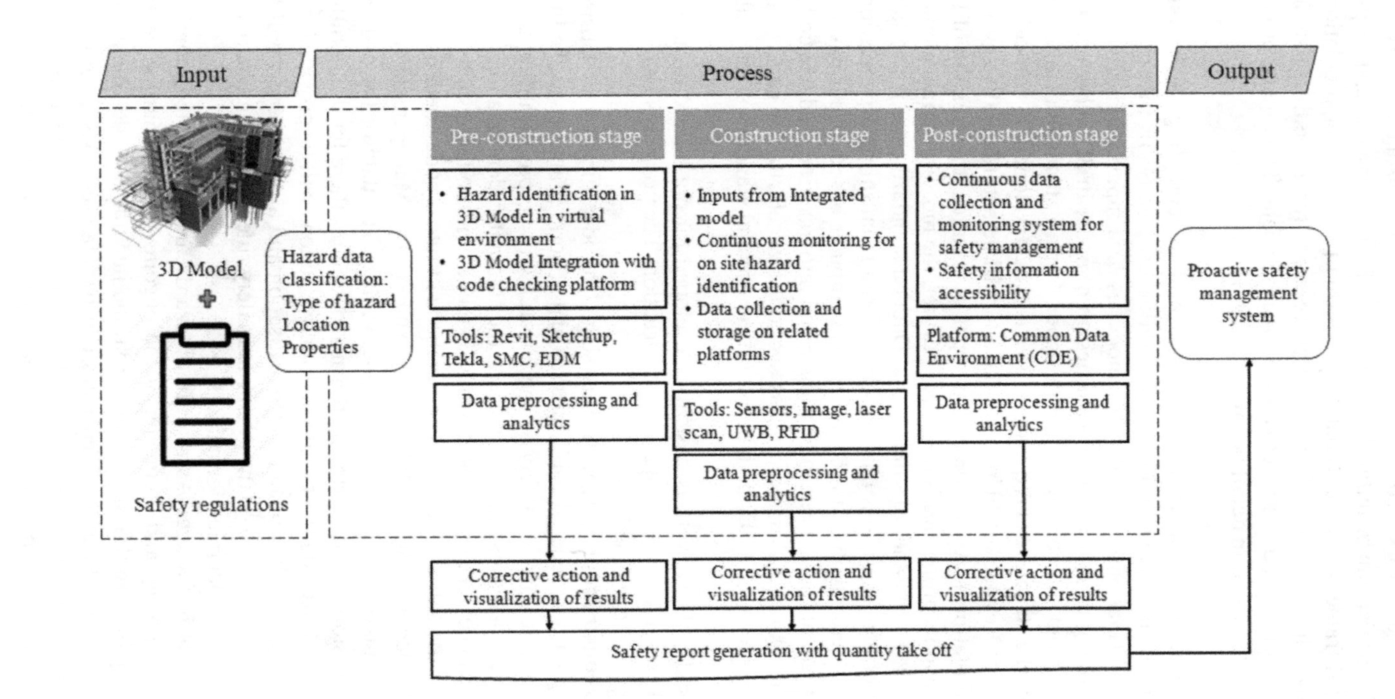

FIGURE 3.5 Framework proposed for proactive safety management.

The proposed framework also provides the possibility of using a variety of tools and data platforms that can be utilised at each stage of safety management. Typically, for the proposed framework input is data in the form of information, regulation, and drawing. It flows through a predefined set of processes to reach the final outcome or a set of outcomes that lead to proactive safety management. The inputs, processes, and outputs are described in detail below.

- **Input:** It is widely acknowledged that product modelling is the only solution for managing building information efficiently. In this process, hazard information is built into the model as data input. Primarily, technical drawing plans are required to design a model in 3D, which can be collected from the contractor and designer. For all stages, a detailed 3D model of the structure is required, which can be developed in a modelling tool such as Revit, SketchUp, or Tekla, and will be automatically saved in the used platform. Also, data on the types of hazards, the location of the hazard, and its properties is required. Further, the most important element of this system is standard safety rules. Data on standard rules is available on the websites of various national and international safety organisations. As the process involved in the system is computer based, paper-based or human readable tools need to be transferred into machine readable codes, which are explained in the process section. Henceforth, for each stage, data in the form of a 3D model of the structure and standard safety rules is necessary.
- **Process:** The framework is basically developed for pre-hazard identification and prevention at each stage of construction. The processes involved at each stage are explained.
  - **Pre-construction stage:** At this stage, the identification of hazards can greatly reduce probable hazards and rework. It also saves the time required to manage safety as the hazards will be pre-identified along with schedules and quantities. Here, the automatic identification of hazards can be done by integrating the 3D model with codified safety rules, where the safety rules would be embedded into a code-checking platform such as SMC, FORNAX, or Express Data Management (EDM) (Narayanswamy, Liu, and Al-Hussein 2019). The data of the codified tools is stored in a selected platform and remains there for use in future projects. Based on safety rules, corrective actions will be provided, and the model will be updated for visualisation and a safety report will be generated. The data on the model and safety rules will be automatically stored in the operated format.
  - **Construction stage:** As real-time hazards are unexpected, continuous monitoring and recording of the ongoing construction processes are required. In this stage, input from the integrated model about hazard scenarios, locations, and rules can be analysed. Also, the design of the construction can be subject to change for various reasons and in this case hazard scenarios can be less known, so the recording and

monitoring of the construction process with techniques such as cameras, radio-frequency identification (RFID), laser scans, and sensors that can locate and identify real-time hazard scenarios are essential. The data generated using selected techniques can be saved in a supported format. The safety manager plays a vital role in capturing and installing the selected technique in the physical environment for on-site hazard identification. Identified hazards can be prevented through integrated safety checking and also allows their prevention through the experience and knowledge of the expert person. In this manner, hazards can be identified, and suitable solutions can be provided to prevent injury. A safety report will be generated at this stage as well.

- **Post-construction stage:** This stage mainly involves the operation and maintenance of the building. The basic advantage of proactive safety checking is that the information on each process can be stored in a system and is available on a single platform for sharing with different actors. BIM has the ability to combine two or more platforms with the Industry Foundation Class (IFC) model. This information and records can be helpful for safety planning in future projects. Once the building is in use, hazards can be predicted based on knowledge and experience and suitable solutions can be provided.

- **Output:** The proposed framework provides construction safety planning for the entire life cycle of the project. A proactive safety system identifies hazards prior to construction, reduces rework, and saves cost and time. Information generated at each stage is stored in the system and it can be used as input data for future projects.

3.4.2 Interaction of Cyber-Physical System Elements in Proactive Safety Management Framework

A cyber-physical system requires two-way communication and synchronisation between virtual models and physical structures to enhance consistency management (Figure 3.6) (Akanmu, Anumba, and Messner 2013). This helps with real-time progress monitoring, construction process quality management and control, tracking of improvements and model changes, knowledge exchange between the design office and the field, as-constructed status reporting, and sustainable practices (Akanmu et al. 2013a). Computational resources are needed to closely combine the cyber and physical worlds in order to achieve two-way coordination, which is referred to as a cyber-physical system strategy. A virtual component for any construction project is the information sets made available using different tools. Cyber components are data sets put on a cloud platform for wider usability and accessibility. The virtual components under the umbrella of a cyber system enable the integration of stakeholder and process management and the synchronisation of data platforms.

By integrating physical and cyber (or informational) elements, the cyber-physical systems approach enables the creation of situation-integrated analytical systems that intelligently respond to dynamic changes in real-world circumstances. In

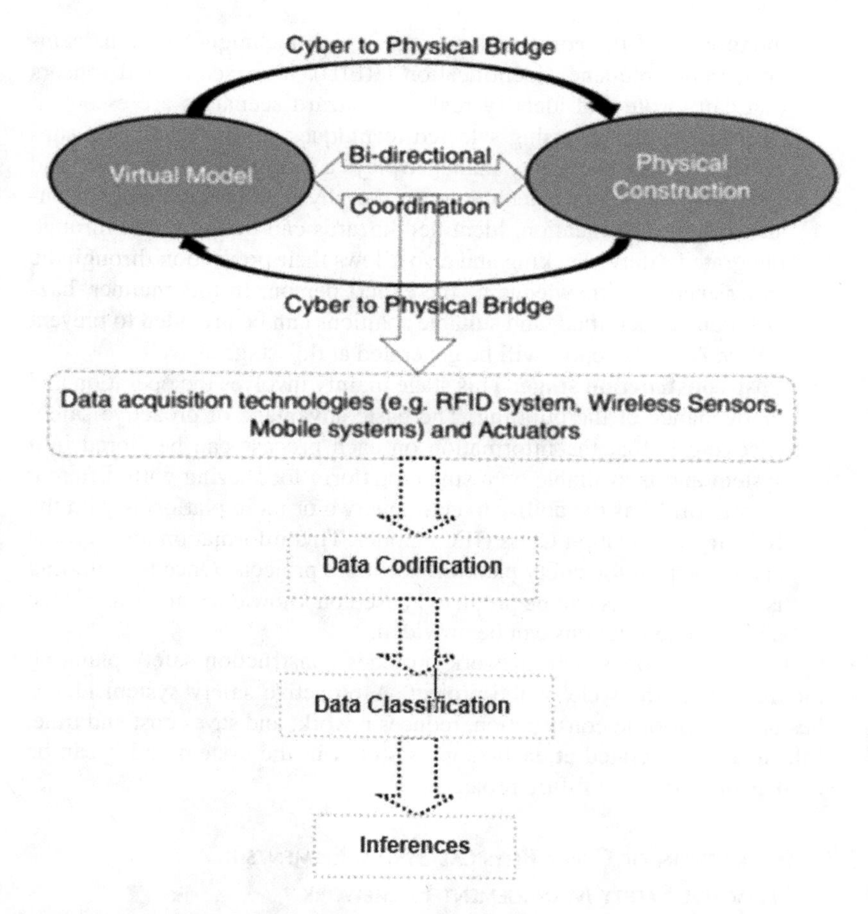

FIGURE 3.6 Components of a cyber-physical system (Akanmu, Anumba, and Messner 2013).

a cyber-physical environment, the cyber component and the physical components closely interact with each other for exchange of information. This process aids in the continuous upgradation of both components.

Many researchers developed the framework for hazard identification at the pre-construction stage, whereas studies based on identifying real-time hazards at the construction stage are still lagging behind. According to previous integration attempts, there is minimal space for the use of virtual models in the building process and facility life cycle as a means of contact between designers in the office and workers on site.

Physical environments, virtual models, a communication network, a database, and portable devices make up the proposed cyber–physical interaction in a proactive safety management process. Virtual models can be used to create both planned and

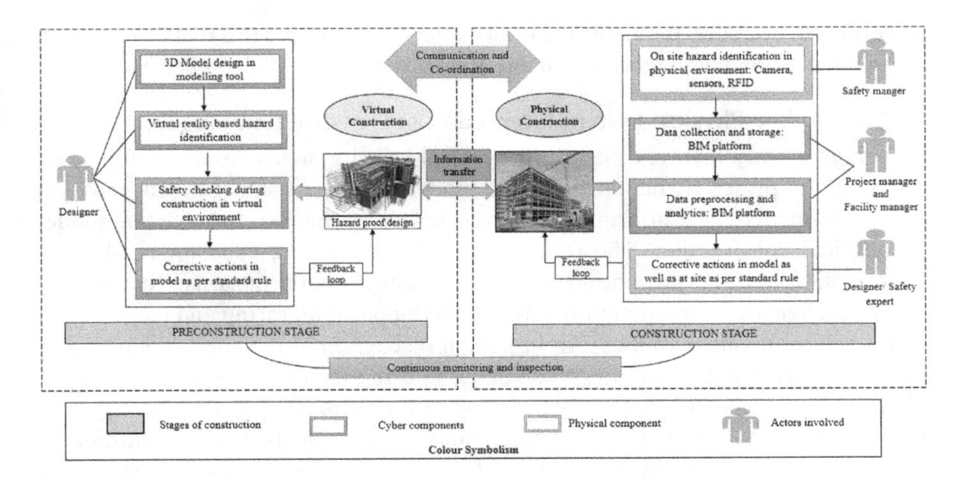

FIGURE 3.7 Interaction of a cyber-physical system in a proactive framework.

constructed construction project scenarios. The ability to compare as-built and as-planned data allows for active monitoring of construction project outcomes. Virtual models are semantic object representations of physical components that can be used to simulate construction processes and store or document building information such as component specifications, position, and so on. The physical construction process, on the other hand, involves numerous construction tasks that must be closely monitored in order to keep the project on time and within budget. Efficient virtual model and physical construction integration should be able to communicate design information, capture and record as-built data, track construction progress, and manage building components in the finished facility. Figure 3.7 illustrates the interaction of the cyber-physical system's elements in the proposed proactive safety management framework.

3.4.2.1 Cyber-Physical System Interaction in Pre-Construction Stage

It is widely acknowledged that product modelling is the only solution to managing building information efficiently. Project life cycle knowledge can be visualised and embedded using virtual models. BIM are virtual models that can store embedded information that can be used during the life cycle of a facility. In this process, hazard information is built into the model as data input. This input mainly contains data on types of hazards, the location of the hazard, and its properties. A required detailed virtual model of a structure can be developed in modelling tools such as Revit, SketchUp, and Tekla, as explained in Section 3.4.1, and the model can be integrated with a safety code-checking platform for rule checking. After checking, the identified hazards are corrected with standard rules and the corrected model with a "hazard proof design" is forwarded to the physical environment. As the model identifies the hazards through design, the key actor involved in the pre-construction process is the designer. Basically, the information is transferred to a real construction site to build as planned in the virtual environment.

3.4.2.2 Cyber-Physical System Interaction in Construction Stage

This stage mainly involves the use of cyber-physical systems front-end functionalities. In the pre-construction stage, the model is updated and retained, ready to build as modelled. The information is transformed in the form of a virtual model for physical construction. The framework is basically developed for pre-hazard identification and the prevention of real-time construction hazards. At the construction stage, the construction site needs continuous monitoring and recording to keep track of on-site activities. An in-person inspection of the site is complicated and time-consuming. However, it can be done using techniques such as cameras, RFID, laser scans, and sensors that can locate and identify real-time hazard scenarios at the location of tagged components, which can then be transmitted to the virtual components in the model. RFID tags can also be used to store and share site information, which is beneficial to facility management. Wireless sensors are typically in charge of information sharing, acting as a link between the cyber and physical systems. These sensors are used in the construction industry to gather information about the facilities being built, as well as the processes and resources involved. The cyber element in the model allows the transmission and sharing of information between sensors, and mobile and fixed devices, for improving bi-directional coordination between the virtual model and the physical construction. The internet, Wi-Fi, and wireless personal area network are examples of communication networks used in the construction industry (comprising UWB and Bluetooth). Using these communication networks, data can be wirelessly transmitted or shared between devices on the construction site, as well as between the construction site and the remote office, allowing the project team to collaborate more effectively. Range, cost, data transfer rate, network topology, and battery life are all factors that affect network selection (Shen, Chen, and Lu 2008). Based on a hazard scenario, the virtual model needs to update with suitable safety rules based on standard regulations. Capturing, installing, and inspecting can be done by the safety manager available on site. Further, the identified data can be exchanged and stored in selected digital platforms through communication and networking. At the construction stage, the data can be collected and stored by the project manager at post-construction; this data can be used by the facility manager for record maintenance purposes. As previously mentioned, the IFC-compliant BIM follows a standard data model by default. Most BIM applications allow you to import and overlay an external file on top of the BIM. The construction crew can use these devices to access virtual models, update construction changes, and submit questions or complaints to the design team regarding specific parts, and more. Basically, at this stage the design of the model is upgraded. Further, a safety report can be generated in the form of visualising real-time hazards and solutions. After updating the model with corrective actions, the same needs to be applied to the physical construction site. It also saves the time required to manage safety as the hazards will be pre-identified along with schedules and quantities.

The benefits of the framework can be the visualisation of the proposed construction site at the design stage, the identification of hazards, training the workers through simulation and visualisation, and continuous inspection and monitoring of the project. Also, the framework allows continuous monitoring and inspection of involved processes throughout the project's life cycle. The proactive safety framework helps to reduce rework, saves time, and automatically reduces the cost involved in reactive safety management practices by managing the cost at the pre-construction level only. The interaction involves various tools that help to store the data generated at each stage and it also allows transferring and sharing the information from the virtual model to the physical environment through cyber components that lead to improved coordination and communication.

3.5 DISCUSSION

Construction safety has emerged as a serious concern for the global construction industry. Despite conventional safety practices, on-site accidents, injuries, and illnesses still occur. Technology is seen as a key tool for improving the health and safety of construction workers in the workplace, as well as ensuring construction safety management in general. The leading concept of Industry 4.0 plays a significant role in managing safety with its digitising and coordinating applicability. To examine the trends of Industry 4.0 in construction safety, a state-of-the-art survey has been conducted in the study. Various representatives of Industry 4.0 have been studied. To provide advanced solutions, this study firstly overviews conventional safety practices and their limitations and then moves towards applying advanced safety over traditional approaches. The conventional approach to safety planning is more reactive. Decisions mainly depend on human input, which is subject to error and is time-consuming as hazard cases may differ from project to project. Major studies on safety management are limited to providing safety at the pre-construction stage; therefore, in this study, the proactive framework is proposed for safety management throughout the life cycle of a construction project. The detailed framework with input, the processes involved, and output is explained along with the tools involved in the system. Further, to make the system more efficient and compatible with physical construction approaches, the cyber-physical element is introduced in the proposed framework. The discovery of a cyber-physical system method in the construction industry has been made possible by rapid advances in ICT. By communicating between the design and construction teams, the cyber-physical approach described in this chapter enables the transition from virtual to physical construction while maintaining consistency between the as-construct and as-planned models. According to the established system, the interaction of cyber-physical elements will help to improve duplex coordination between virtual models and physical construction. Via scene reconstruction design, data collection and storage, data communication, and data processing modules, the device establishes design and safety data synchronisation mapping between virtual and physical construction sites. It also proposes using sensors, RFID, and images to capture real-time on-site hazard information, and shows the actors involved in the interaction process. The proposed framework provides options of available digital techniques that can be used in the process as they apply to individual projects. The framework mainly shows how the data is transferred from one platform to another. Each stage involves data input that is stored in a suitable platform with the competencies of BIM. Drift in data safety management is seen as a new and effective paradigm for addressing the challenges that traditional practices face, as well as for achieving smart safety management in the age of big data. The data collected in the process is saved indefinitely and can be used in future projects. In the information age, particularly in the era of big data, safety management with data drift is an evolving approach to help, direct, and develop safety data management and achieve smart safety management. As a result, it can be concluded that investing in innovations for safety management with data drift strengthens construction site safety.

REFERENCES

Akanmu, Abiola, Chimay Anumba, and John Messner. "Scenarios for cyber-physical systems integration in construction". *Journal of Information Technology in Construction (ITcon)* 18, no. 12 (2013): 240–260.

Alaloul, Wesam Salah, M. S. Liew, Noor Amila Wan Abdullah Zawawi, and Ickx Baldwin Kennedy. "Industrial Revolution 4.0 in the construction industry: Challenges and opportunities for stakeholders". *Ain Shams Engineering Journal* 11, no. 1 (2020): 225–230.

Arslan, Muhammad, Christophe Cruz, and Dominique Ginhac. "Visualizing intrusions in dynamic building environments for worker safety". *Safety Science* 120 (2019): 428–446.

Barata, João, and Paulo Rupino da Cunha. "Safety is the new black: The increasing role of wearables in occupational health and safety in construction". In *International Conference on Business Information Systems*, pp. 526–537. Springer, Cham, 2019.

Chantawit, Damrong, Bonaventura H. W. Hadikusumo, Chotchai Charoenngam, and Steve Rowlinson. "4DCAD-Safety: Visualizing project scheduling and safety planning". *Construction Innovation* (2005).

Ganah, Abdulkadir, and Godfaurd A. John. "Integrating building information modeling and health and safety for onsite construction". *Safety and Health at Work* 6, no. 1 (2015): 39–45.

Giretti, Alberto, Alessandro Carbonari, Berardo Naticchia, and Mario DeGrassi. "Design and first development of an automated real-time safety management system for construction sites". *Journal of Civil Engineering and Management* 15, no. 4 (2009): 325–336.

Hire, S., Ruikar, K., Sandbhor, S., and Amarnath, C. B. "A critical review on BIM for construction safety management". In *Proceedings of PMI India Research & Academic Virtual Conference*, 3–6 March, 2021a.

Hire, Shalaka, Sayali Sandbhor, and Kirti Ruikar. "Bibliometric survey for adoption of building information modeling (BIM) in construction industry–a safety perspective". *Archives of Computational Methods in Engineering* (2021b): 1–15.

Hongling, Guo, Yu Yantao, Zhang Weisheng, and Li Yan. "BIM and safety rules based automated identification of unsafe design factors in construction". *Procedia Engineering* 164 (2016): 467–472.

Hossain, Md Mehrab, and Shakil Ahmed. "Developing an automated safety checking system using BIM: A case study in the Bangladeshi construction industry". *International Journal of Construction Management* (2019): 1–19. https://doi.org/10.1080/15623599.2019.1686833.

Jiang, Weiguang, Lieyun Ding, and Cheng Zhou. "Cyber physical system for safety management in smart construction site". *Engineering, Construction and Architectural Management* (2020).

Kanchana, S., P. Sivaprakash, and Sebastian Joseph. "Studies on labour safety in construction sites". *The Scientific World Journal* (2015). https://doi.org/10.1155/2015/590810.

Kim, Kyungki, Yong Cho, and Sijie Zhang. "Integrating work sequences and temporary structures into safety planning: Automated scaffolding-related safety hazard identification and prevention in BIM". *Automation in Construction* 70 (2016): 128–142.

Lu, Yang. "Industry 4.0: A survey on technologies, applications and open research issues". *Journal of Industrial Information Integration* 6 (2017): 1–10.

Lu, Ying, Qiming Li, Zhipeng Zhou, and Yongliang Deng. "Ontology-based knowledge modeling for automated construction safety checking". *Safety Science* 79 (2015): 11–18.

Narayanswamy, H., H. Liu, and M. Al-Hussein. "BIM-based Automated Design Checking for Building Permit in the Light-Frame Building Industry". In *ISARC. Proceedings of the International Symposium on Automation and Robotics in Construction*, 36, pp. 1042–1049. IAARC Publications, 2019.

Naticchia, Berardo, Massimo Vaccarini, and Alessandro Carbonari. "A monitoring system for real-time interference control on large construction sites". *Automation in Construction* 29 (2013): 148–160.

Oesterreich, Thuy Duong, and Frank Teuteberg. "Understanding the implications of digitisation and automation in the context of Industry 4.0: A triangulation approach and elements of a research agenda for the construction industry". *Computers in industry* 83 (2016): 121–139.

OSHA. 2011. "Construction focus four: Outreach training packet", no. April: 22. https://www.osha.gov/dte/outreach/construction/focus_four/constrfocusfour_introduction.pdf.

Osunsanmi, Temidayo O., Clinton Aigbavboa, and Ayodeji Oke. "Construction 4.0: The future of the construction industry in South Africa". *International Journal of Civil and Environmental Engineering* 12, no. 3 (2018): 206–212.

Patel, Dilipkumar Arvindkumar, and Kumar Neeraj Jha. "An estimate of fatal accidents in Indian construction". In *Proceedings of the 32nd Annual ARCOM Conference*, vol. 1, pp. 577–586. 2016.

Romero, David, Sandra Mattsson, Åsa Fast-Berglund, Thorsten Wuest, Dominic Gorecky, and Johan Stahre. "Digitalizing occupational health, safety and productivity for the operator 4.0". In *IFIP International Conference on Advances in Production Management Systems*, pp. 473–481. Springer, Cham, 2018.

Schober, Kai-Stefan and Philipp Hoff. 2016. "Digitization in the construction industry". *ROLAND BERGER GmbH*, 16. https://www.rolandberger.com/publications/publication_pdf/tab_digitization_construction_industry_e_final.pdf.

Shen, Xuesong, Wu Cheng, and Ming Lu. "Wireless sensor networks for resources tracking at building construction sites". *Tsinghua Science and Technology* 13, no. S1 (2008): 78–83.

Vacanas, Yiannis, Kyriacos Themistocleous, Athos Agapiou, and Diofantos Hadjimitsis. "Building Information Modelling (BIM) and Unmanned Aerial Vehicle (UAV) technologies in infrastructure construction project management and delay and disruption analysis". In *Third International Conference on Remote Sensing and Geoinformation of the Environment* (RSCy2015), vol. 9535, p. 95350C. International Society for Optics and Photonics, 2015.

Yulistianingsih, Trijeti, Lidia Sarah Fairyo, Anik Setyo Wahyuningsih, et al. "Penerapan Sistem Pengendalian Keselamatan Dan Kesehatan Kerja Pada Pelaksanaan Konstruksi (Studi Kasus: Lanjutan Pembangunan Fasilitas Pelabuhan Laut Manado)". *Konstruksia – Universitas Muhammadiyah Jakarta* 2, no. 1 (2012): 17–25.

Zhang, Sijie, Jochen Teizer, Jin-Kook Lee, Charles M. Eastman, and Manu Venugopal. "Building information modeling (BIM) and safety: Automatic safety checking of construction models and schedules". *Automation in Construction* 29 (2013): 183–195.

Zhou, Wei, Jennifer Whyte, and Rafael Sacks. "Construction safety and digital design: A review". *Automation in Construction* 22 (2012): 102–111.

Zhou, Zhipeng, Javier Irizarry, and Qiming Li. "Applying advanced technology to improve safety management in the construction industry: A literature review". *Construction Management and Economics* 31, no. 6 (2013): 606–622.

4 Machinery Fault Detection using Artificial Intelligence in Industry 4.0

Pooja Kamat, Sıtkı Akıncıoğlu, and Rekha Sugandhi

CONTENTS

4.1 THE ERA OF INDUSTRY 4.0

Since the 1800s, the world has been revolutionised by ground-breaking industrial technologies, from steam engines for generating power, assembly lines for mass manufacturing, and finally the computer for making the entire process swift and efficient (Sartal et al. 2020). We are currently witnessing the Industry 4.0 wave, which is disrupting all industry verticals. Industry 4.0 is mainly derived from the industrial internet of things (IIoT) and cyber-physical systems, and encompasses

DOI: 10.1201/9781003200857-4

smart, automated systems that are driven by algorithmic-based techniques. These techniques backed by computer-based intelligence are used to track, control, and automate tangible objects such as equipment, robots, and vehicles. Everything in the supply chain becomes "smart" as a result of Industry 4.0. It integrates with back-end systems such as enterprise resource planning (ERP) to provide businesses with unparalleled insight and power. We are rapidly progressing towards witnessing the next Industrial Revolution, termed Industry X.0, which aims to realise the digital value in industrial sectors using a variety of next-generation applications. Figure 4.1 depicts the progress of the industrial revolutions towards Industry X.0.

4.1.1 Sustainable Smart Manufacturing under Industry 4.0

Globally, sustainable manufacturing has always been of great importance to the manufacturing industry. The chief vision of sustainable manufacturing is to realise manufacturing processes and product development that have the least impact on the environment. To make this vision a reality, extensive research in terms of resources and energy consumption is necessary. Smart manufacturing is the key to achieving sustainable manufacturing. It is the amalgamation of intelligent data-driven decision-making and manufacturing processes. Smart sensors are used to extract the health-monitoring data of machinery, and artificial intelligence (AI) analytical tools further process the data to make critical decisions. One such contribution of smart manufacturing is predictive maintenance (PdM) which is explained in the next section.

4.1.2 Introduction and Evolution of Predictive Maintenance

> Maintenance is a strategic concern when developing and manufacturing a product – and for a good reason
>
> **– Anonymous**

Timely maintenance is the primary goal for the manufacturing industry to achieve optimum production at minimum cost. An alarming report in Wollenhaupt (2016) mentions that ineffective maintenance strategies can reduce a machinery plant's productive capacity by 5–10%. Manufacturers deal with different levels of maintenance strategies according to their organisation's capabilities. The four basic levels are depicted in Figure 4.2 (Zonta et al. 2020).

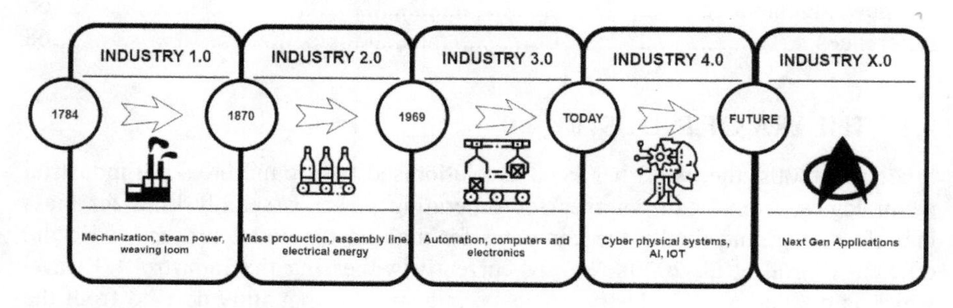

FIGURE 4.1 Progressing towards Industry X.0.

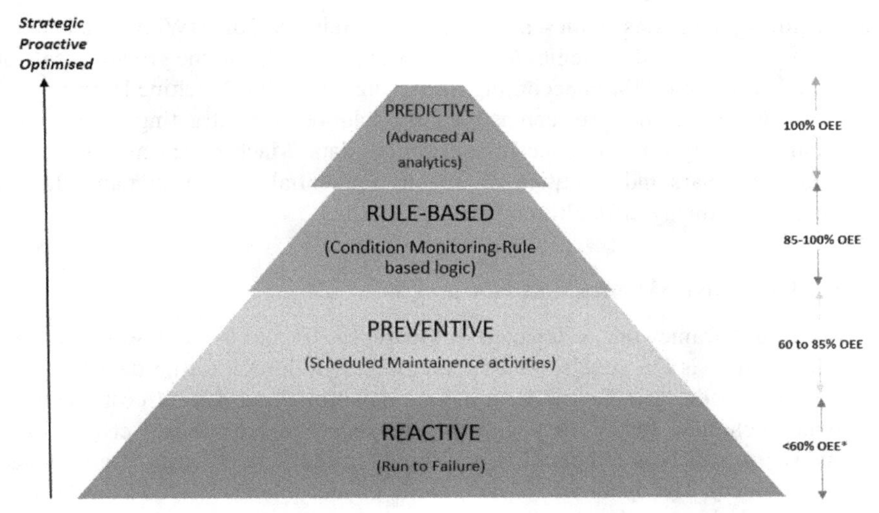

FIGURE 4.2 Levels of maintenance strategies.

Reactive maintenance is basically machinery that is run without carrying out maintenance until failure, e.g. changing a bulb after it is broken. Although it imposes less of a maintenance burden, the overall life of the machinery is compromised. Preventive maintenance is periodic inspections and maintenance but generally without any insights, for example, carrying out the scheduled maintenance of a vehicle every three months even though it is new and running perfectly fine. Considerable machinery effectiveness is obtained via preventive maintenance. However, preventive maintenance can lead to unnecessary wastage of maintenance resources. Rule-based maintenance, also known as condition monitoring, continuously collects machinery data on which predefined rule alerts are set. These predefined rule alerts often lead to false alarms or late alarms as they are not updated as per the dynamics of the operating environment. Finally, predictive maintenance unleashes the power of artificial intelligence in making intelligent maintenance decisions using machine learning (ML) and deep learning algorithms on historical machinery data. Optimum equipment effectiveness is achieved via predictive maintenance. One of the initial steps in predictive maintenance is the technique of anomaly detection or fault detection in the working of the machinery. Early detection of faults using AI tools can assist the machinery supervisor in strategising maintenance activities (Kamat and Sugandhi 2020a).

4.2 ARTIFICIAL INTELLIGENCE–EMPOWERED PREDICTIVE MAINTENANCE

In recent years, artificial intelligence–based solutions for performing predictive maintenance operations have become increasingly popular across a variety of manufacturing industries (Janiesch, Zschech, and Heinrich 2021). With the advancement of big data–related strategies (e.g. sensors and the IoT) and the growing scale of big data, data-driven PdM is becoming increasingly popular. Machine learning and deep learning approaches are seen as effective solutions for extracting useful information and making informed decisions from big data. Machine learning is widely explored by scholars and practitioners alike as a potential solution to many old and current manufacturing difficulties (Wuest et al. 2016).

4.2.1 Predictive Maintenance Framework

A typical PdM framework is depicted in Figure 4.3. The first step towards predictive maintenance is the acquisition of the correct health-monitoring data from the machine. Given the growing need to reduce downtime and related costs, PdM is a popular technique for dealing with maintenance concerns (Susto et al. 2015). Various sensors such as temperature, vibration, current, and acoustic are mounted on the machinery. The operational data is recorded and either stored on a local server or transmitted via a wireless medium to the cloud server. The recorded data is very "raw" in format and initially requires certain preprocessing techniques before it is sent to the prediction models. During preprocessing, data augmentation is needed to replace missing sensor values. Data imputation and encoding are carried out to convert the data into a representable format.

Next, time domain, frequency domain, and time–frequency domain analysis are various signal processing techniques that can convert the raw vibration signals into more meaningful information. The processed information is then divided into two sets: training and testing. The typical ratio used in this split is either 70–30% or 80–20%. The model is trained on the training set, from which it learns feature correlations. The model results are validated on the testing set. Next, the processed data is fed to the AI engine for making intelligent decisions. This is divided into two stages: fault diagnosis and fault prognosis.

Fault diagnosis follows a classification approach wherein the model can classify whether a fault is present or not (Yu et al. 2020). Dimensionality reduction is carried out initially to get a primed data set. Further data denoising and feature extraction are carried out so that the model is trained only on the essential features contributing to the fault. The trained model's classification accuracy can be further improved by optimising the algorithm threshold. This is known as hyperparameter tuning. The model performance is then authenticated on the testing set. Finally, on deployment the model can generate alerts when a new anomaly or fault is detected in the working of the machinery.

Fault prognosis follows a regression approach wherein the anomaly detected in the diagnosis stage can be valuable input for estimating the remaining useful life (RUL) of the machinery (Jasiulewicz-Kaczmarek and Gola 2019). RUL estimation involves training regression-based models to understand the sequential dependencies using the timestamps of the recorded data. This pattern analysis helps the model estimate how many time cycles are left before the machinery shuts down if no maintenance is carried out.

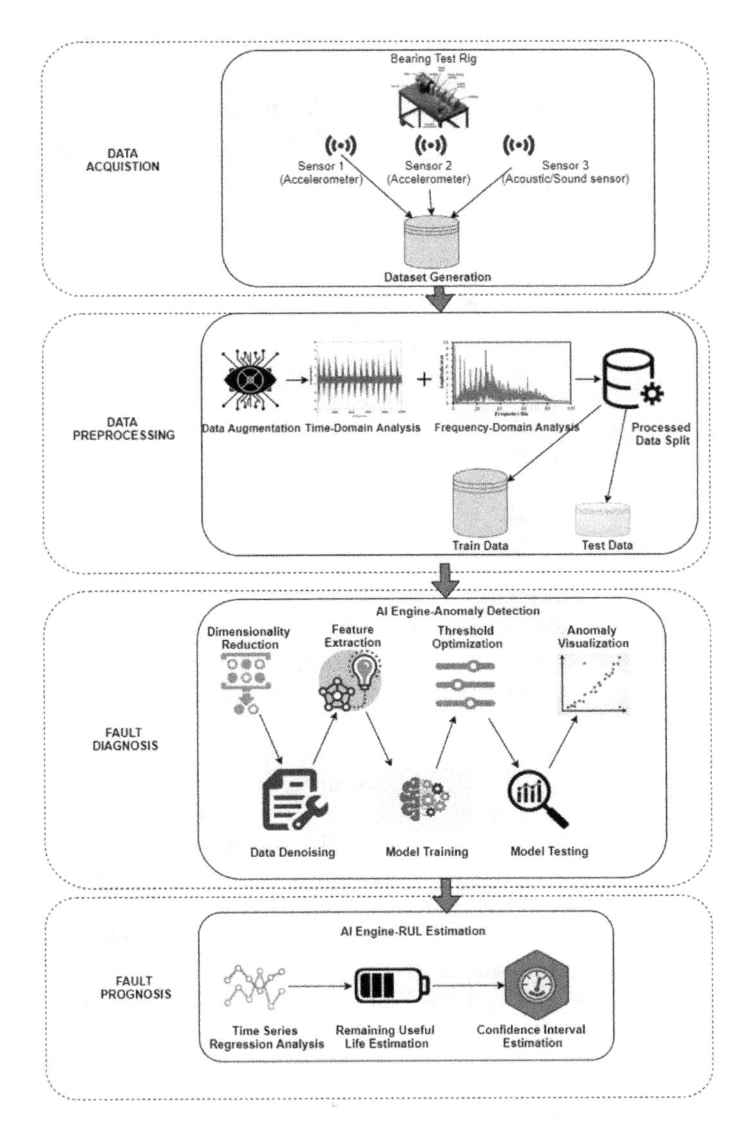

FIGURE 4.3 AI-enabled predictive maintenance framework.

4.2.2 Literature Survey: Fault Detection and Diagnosis

Data collection, feature extraction, and anomaly detection are the three main steps of fault detection and diagnosis (FDD). In FDD systems, data acquisition is critical for obtaining the signal representing the physical state of the machinery portion. Acoustic emissions (AE) signals, vibration data, pressure data, oil analysis, and heat analysis are some of the condition monitoring methods employed in this process (X. Chen et al. 2018).

TABLE 4.1
Recent Work in FDD Based on Machinery, Algorithm, and Sensors Used

Year	Author	Machinery	Algorithm	Sensors		
				Vibration	Acoustic	Current
2020	Yuan et al. (2020)	Rolling bearing	Support vector machine	✓		
2019	Tian et al. (2020)	Inter-shaft bearing	Random forest classifier		✓	
2020	Azamfar et al. (2020)	Gearbox	Convolutional neural network			✓
2019	Dhanraj, Sugumaran, and Joshuva (2021)	Wind turbine blade	k-Nearest neighbour	✓		
2020	Ou et al. (2021)	CNC milling	Stacked denoising autoencoder			✓

For effective feature extraction, signal processing techniques are widely used. Chen et al. proposed an elaborate bibliometric review of the steps involved in FDD for rotating machinery (J. Chen et al. 2020). Some popular techniques are spectral kurtosis (SK) and kurtogram methods (Zhang and Randall 2009), wavelet-based methods (Tao et al. 2020), autoregressive methods (Tao et al. 2020), and empirical mode decomposition (EMD) methods (Shi et al. 2020).

Finally, there are three ways to detect anomalies: model driven–based, data/signal driven–based, and knowledge driven–based techniques (Eren, Ince, and Kiranyaz 2019; Ellefsen et al. 2019). Researchers have paid more attention to signal-based and data-driven methods in modern industry because they provide high diagnostic precision (H. Wang et al. 2019) and do not require observational measurement of physical parameters. Saufi et al. present an extensive survey of the machine learning and deep learning techniques used in the FDD of rotating machinery (Saufi et al. 2019). Table 4.1 presents some of the recent work on FDD, highlighting the machinery, algorithms, and sensors used.

4.3 CASE STUDY ON FAULT DETECTION IN BEARING MACHINERY

Bearings are used in a range of automotive and manufacturing applications, and are one of the most common machinery components that a supervisor monitors to assess the moving machinery's health. These machines are typically used in mass manufacturing and are subjected to harsh environmental pressures such as high air temperature and high moisture levels for extended periods. The resultant faults from these conditions can lead to massive machinery downtime and increased repair costs (Kamat and Sugandhi 2020b). Figure 4.4 shows the different kinds of faults that occur in a typical roller bearing.

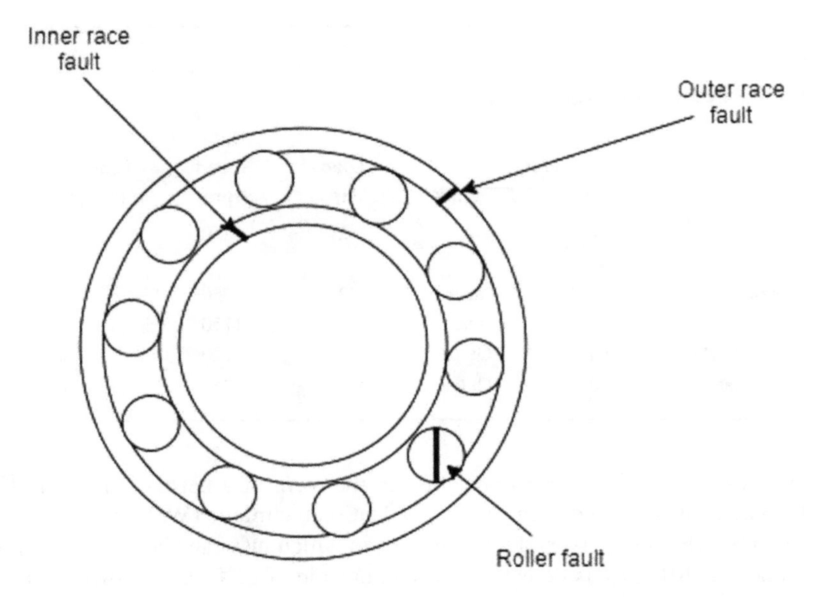

FIGURE 4.4 Structure of a roller bearing representing various types of faults.

Machine learning approaches have become more prominent in the identification of bearing faults due to their numerous advantages. They provide dimensionality reduction, noise removal from data, pattern analysis in sensor data, etc. (Zebari et al. 2020). In this case study, the authors present a support vector machine (SVM) application for bearing fault detection on the Case Western Reserve University (CWRU) bearing data set.

4.3.1 Data Set Description

A CWRU bearing data set was used in this research study ("Case Western Reserve University Bearing Data Center Website | Bearing Data Center" 2021). Matlab data files were translated to comma-separated values (.csv) using the writable function. A standard normal baseline with 12k drive-end (DE) and fan-end (FE) bearing fault data of inner race 0.021" diameter and standard normal baseline with 12k drive-end and fan-end bearing fault data of outer race 0.021" diameter were merged and used in the analysis. All of the data was collected at a motor speed of about 1750 rpm. Table 4.2 describes this case study's data set.

4.3.2 Algorithm Description

Support vector machines are a class of machine learning techniques that have been devised to solve both classification and regression-based problems. First and foremost, their scientific roots have been thoroughly investigated, and they provide a convex optimisation technique ensuring that global optimum is achieved. In addition,

TABLE 4.2
Data Set Used in the Study

	Fault diameter (in.)	Fault race	Load (Hp)	Motor speed (rpm)	Length of data file (No. data points)
48k Normal baseline data	NA	NA	2	1750	485063
12k Drive-end data	0.021	Inner	2	1750	121846
12k Fan-end data	0.021	Inner	2	1750	120984
12k Drive-end data	0.021	Outer	2	1750	122281
12k Fan-end data	0.021	Outer	2	1750	120617

they may use a non-linear transformation in the form of a kernel that even allows SVMs to be considered a dimensionality reduction technique (W. Wang et al. 2003). One-class SVMs were designed for scenarios in which just one class has been established and the difficulty is detecting outliers outside of it. This is known as novelty detection, and it refers to the automatic detection of unusual or unexpected events (Pimentel et al. 2014). As a result, one-class SVMs are widely utilised in anomaly identification. Figure 4.5 depicts the architecture of a one-class SVM algorithm.

4.3.3 Methodology

Figure 4.6 presents the methodology used in this case study. The CWRU bearing data files are in .mat format. They were converted to .csv files using the writable function available on Matlab. The data set was then imported using Colab Jupyter Notebook. Certain preprocessing techniques were carried out, such as annotating normal data with 0 and faulty data with 1, and replacing missing values. The data

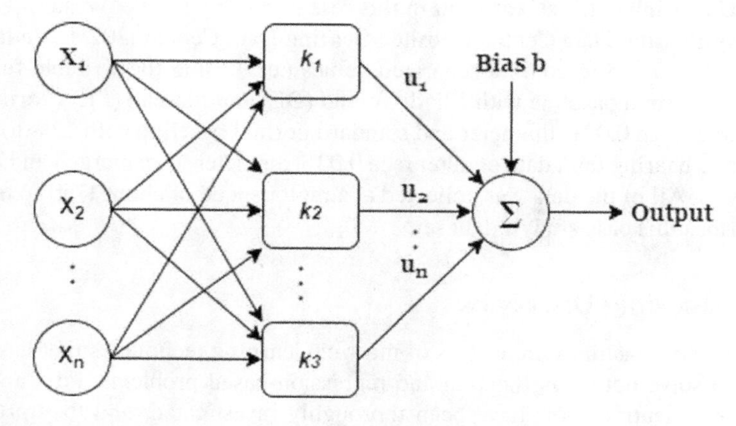

FIGURE 4.5 Architecture of support vector machine.

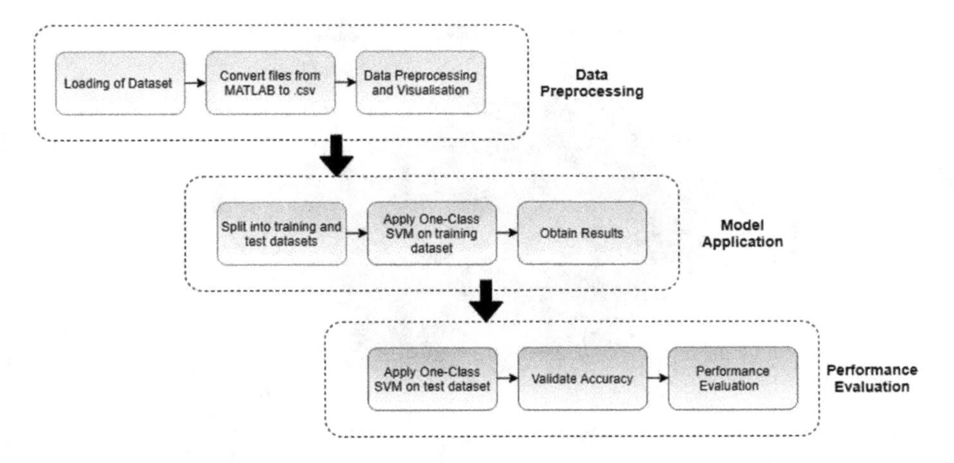

FIGURE 4.6 Methodology used in the case study.

was then visualised using a scatter plot to understand the spread of the data on the basis of the DE and FE feature sets.

The data was then split into training and testing data sets using the sklearn's train_test split selection function. The one-class SVM model was applied for fault detection separately on the DE and FE training samples, and the time series plot was visualised using a matplotlib package. The algorithm's accuracy was validated on the DE and FE testing samples, and the performance was evaluated. The accuracy score and count of anomalies detected were measured at the end.

4.3.4 Results and Analysis

4.3.4.1 Data Visualisation

Figure 4.7a–c depicts a scatter plot of a normal baseline bearing, a faulty inner race fault 0.021", and a faulty outer race fault 0.021" bearing data. The scatter plot is plotted based on the fan-end and drive-end vibration signals during the normal and faulty operation of machinery. The scatter plot helps to visualise the dynamicity in the vibration signals. From Figure 4.7, it can be observed that the vibration data in the normal baseline operational state of the bearing machinery is stable as compared to the vibrational data in the faulty operational state of the machinery.

4.3.4.2 Model Application

The one-class SVM technique is applied for anomaly detection on both inner and outer race bearing data. Anomaly detection is carried out on DE and FE accelerometer vibration data separately. The data is divided into training and testing data sets, and the algorithm is used to validate both sets. The training data set represents the initiation of anomaly points from a normal to an abnormal state in the machinery. The testing data set represents the anomaly points when the vibration signals are at elevated peaks.

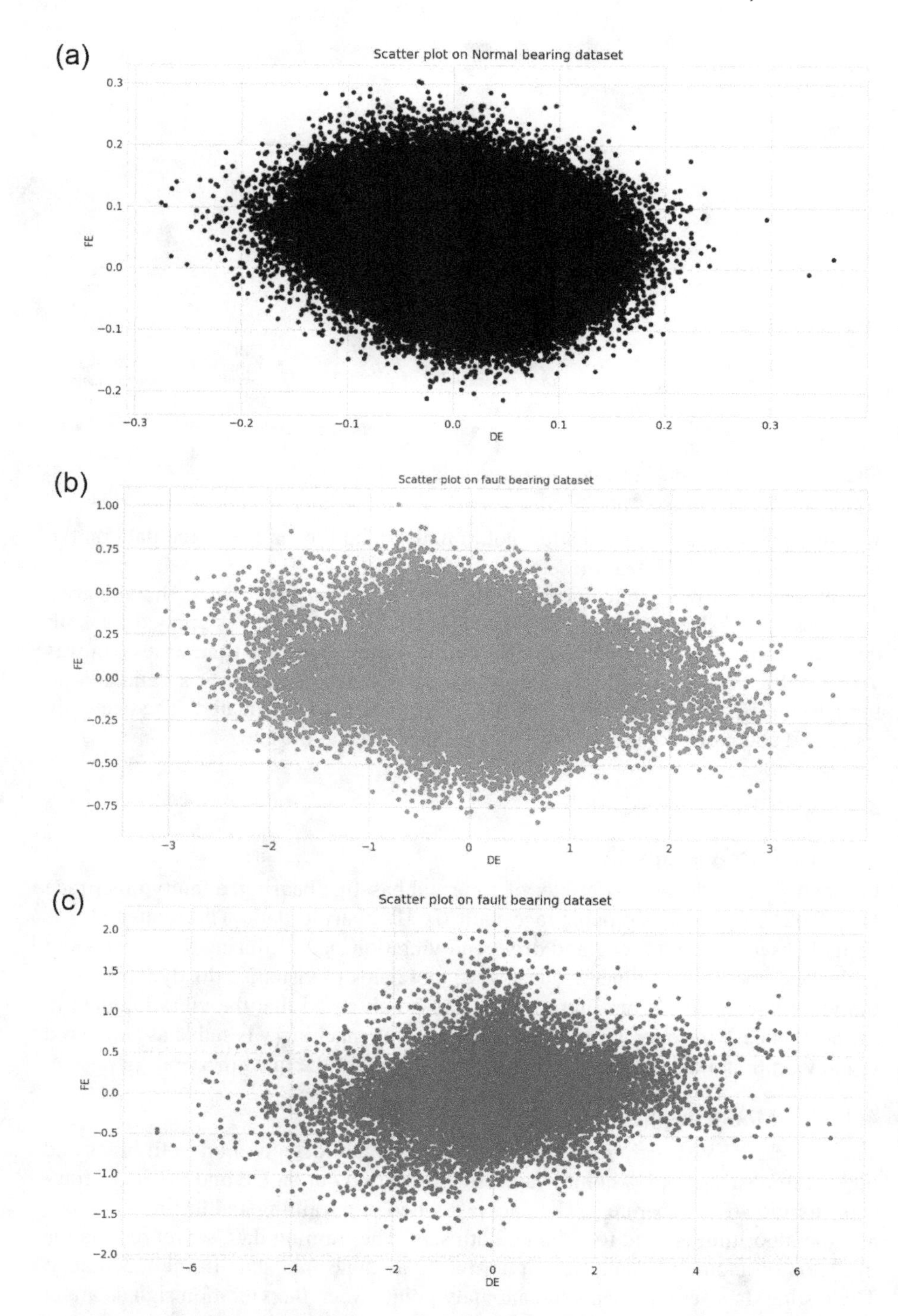

FIGURE 4.7 Scatter plot of normal and faulty bearing data: (a) normal baseline data; (b) faulty inner race 0.021" data; and (c) faulty outer race 0.021" data.

4.3.4.2.1 One-Class SVM for Anomaly Detection on Inner Race 0.021" Fault

Figures 4.8 and 4.9 depict the application of the one-class SVM algorithm on the normal and faulty bearing data on inner race 0.021".

Figure 4.8a and b depicts the anomalies detected by one-class SVM on a training and test data set based on the drive-end accelerometer vibration data. The X-axis represents the inner race vibration data at the drive-end, and the Y-axis represents the vibration signals' anomaly threshold. The black markers represent the anomalous data points.

Figure 4.9a and b depicts the anomalies detected by one-class SVM on a training and test data set based on the fan-end accelerometer vibration data. The X-axis represents the inner race vibration data at the fan-end, and the Y-axis represents the vibration signals' anomaly threshold. The black markers represent the anomalous data points.

4.3.4.2.2 One-Class SVM for Anomaly Detection on Outer Race 0.021" Fault

Figures 4.10 and 4.11 depict the application of the one-class SVM algorithm on the normal and faulty bearing data on outer race 0.021". Similar to the inner race fault classification, the X-axis represents the outer race vibration data at drive-end and fan-end and the Y-axis represents the anomaly threshold in the vibration signals. The black markers represent the anomalous data points.

Figure 4.10a and b depicts the anomalies detected by one-class SVM on training and test data set on the basis of the drive-end accelerometer vibration data. Figure 4.11a

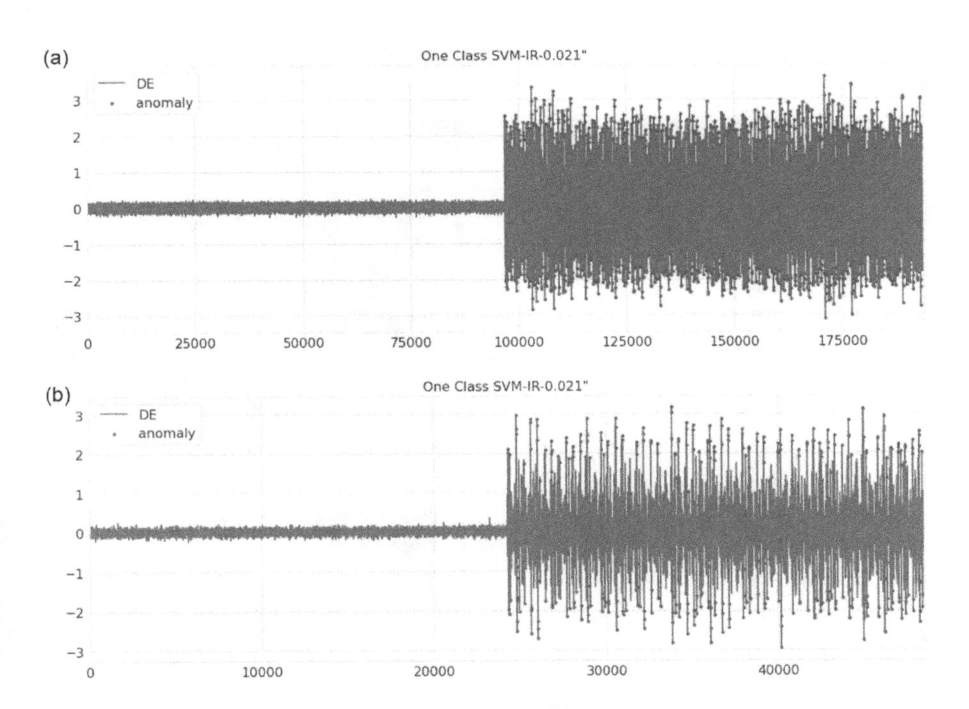

FIGURE 4.8 Anomaly detection by one-SVM on drive-end (DE) accelerometer data of inner race 0.021" fault. (a) Anomalies on the training set; and (b) anomalies on the test set.

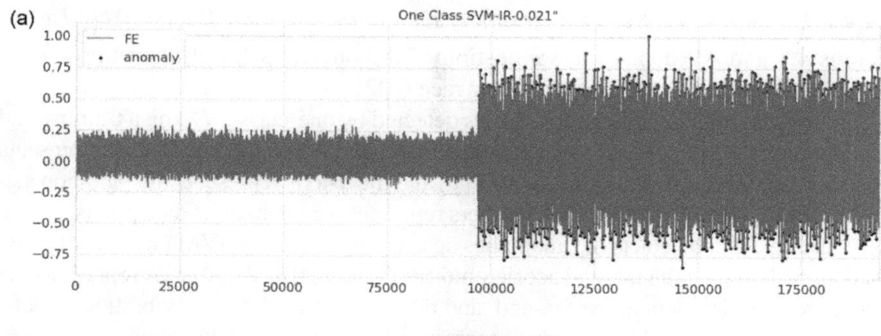

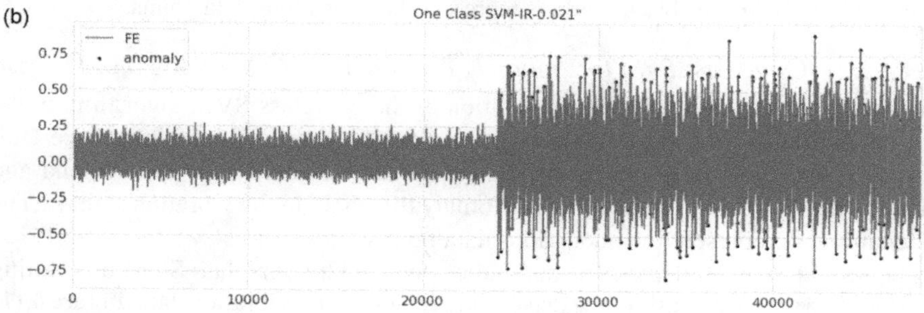

FIGURE 4.9 Anomaly detection by one-SVM on fan-end (FE) accelerometer data of inner race 0.021" fault. (a) Anomalies on the training set; and (b) anomalies on the test set.

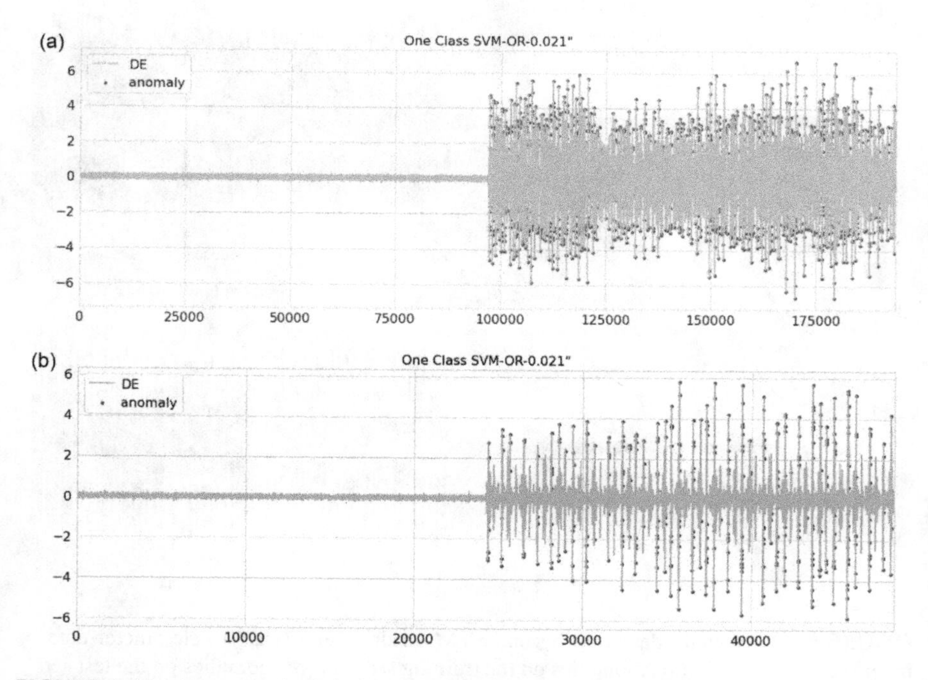

FIGURE 4.10 Anomaly detection by one-SVM on drive-end (DE) accelerometer data of outer race 0.021" fault. (a) Anomalies on the training set and (b) anomalies on the test set.

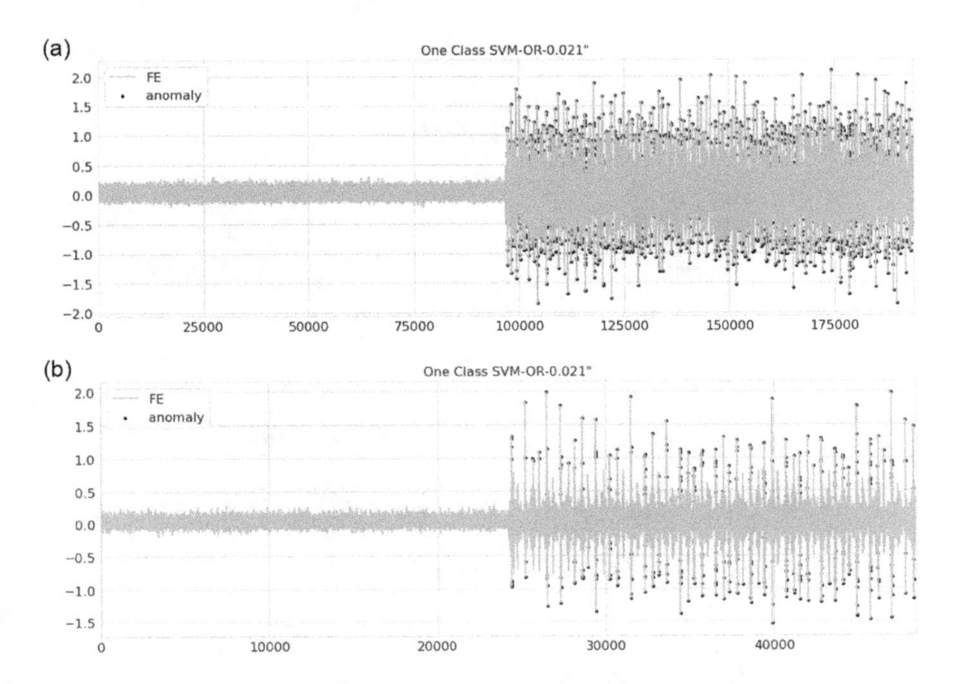

FIGURE 4.11 Testing and training data set for anomaly detection by one-SVM on fan-end (FE) accelerometer data of outer race 0.021" fault. (a) Anomalies on the training set and (b) anomalies on the test set.

and b depicts the anomalies detected by one-class SVM on the training and test data set on the basis of the fan-end accelerometer vibration data.

4.3.4.3 Performance Evaluation

Using both inner and outer race bearing data, Table 4.3 shows the performance details of the one-class SVM method for bearing defect identification. The algorithm exhibited an average accuracy of 60% and 65% for inner race and outer race fault detection, respectively. The SVM displayed an average performance in comparison to other advanced deep learning techniques. The authors propose to combine SVM with pre-trained models such as Alex-Net to improve model accuracy.

TABLE 4.3
Performance Table of One-Class SVM Algorithm

	Inner race 0.021"	Outer race 0.021"
Accuracy	0.60	0.65
Count of anomalies	482	484

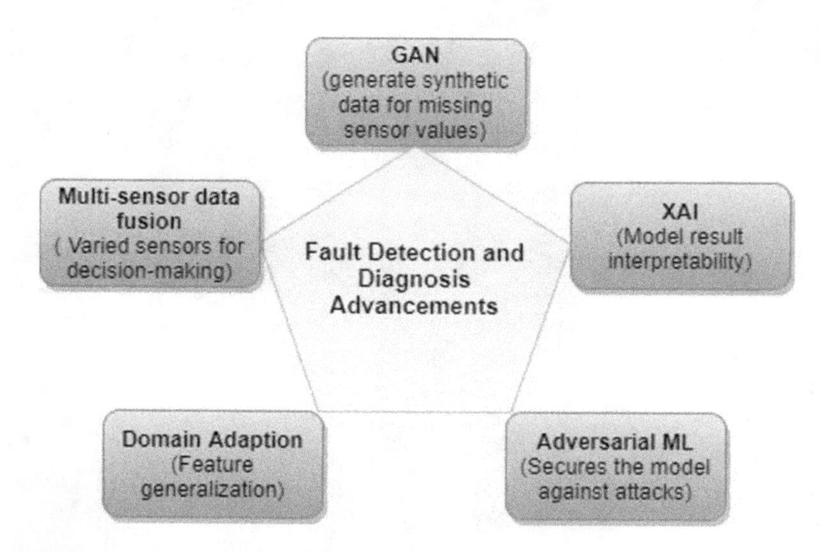

FIGURE 4.12 Future AI-led advancements in FDD.

4.4 FUTURE DIRECTIONS

Fault detection and diagnosis using artificial intelligence have undergone major evolutions over the past few years. AI-led techniques such as generative adversarial networks, domain adaptation, explainable AI, digital twin and adversarial ML, as seen in Figure 4.12, have further enhanced the capabilities of these FDD systems. Table 4.4 highlights some of the future directions in this domain.

4.5 CONCLUSIONS

FDD frameworks are essential in a rotating machinery set-up to maximise machine uptime and avoid unexpected breakdowns. AI models will remain powerful and appealing for use in FDD systems as the big data boom continues to evolve at a rapid pace. AI-led FDD systems can assist industry personnel in realising the goal of sustainable manufacturing. Intelligent decisions regarding fault diagnostics based on the predictive capabilities of AI models can alert the machinery supervisor of a possible machinery breakdown in the future. This chapter discussed the predictive maintenance framework, related work, and a case study on FDD systems. The authors propose to extend this work by applying AI-advancement techniques for fault detection on the CWRU data set in the future.

ACKNOWLEDGEMENT

The authors are grateful to Case Western Reserve University for making the bearing vibration experimental data available on their website free of charge.

TABLE 4.4
Challenges, AI Advancements, and Solutions in FDD

Sr No.	Challenges in the current FDD system	Advancement	Solution	References
1.	Missing sensor values due to malfunctioning of the data acquisition system.	Generative adversarial networks (GAN)	GANs can be used to generate synthetic data in place of missing sensor values.	Mao et al. (2019), Jiang et al. (2019)
2.	Varied machinery operating conditions and set-ups can result in the collection of fewer training samples. This leads to imbalanced data distribution across training and testing sets.	Domain adaptation (DA)	Domain adaptation is a technique by which we train the model of the source domain having a larger data set and later use the pre-trained model to test on the target domain having a smaller data set.	X. Li et al. (2019), Mao, Sun, and Wang (2021)
3.	FDD model predictions often lack the reason behind the predictions made.	Explainable AI (XAI)	XAI techniques decrypt the black box of these model predictions and provide interpretable justifications.	Chatterjee and Dethlefs (2020), Krishnamurthy et al. (2020)
4.	FDD systems might be susceptible to adversarial attacks, which can compromise the system's credibility.	Adversarial machine learning (AML)	Adversarial ML models are trained using pre-designed adversarial samples to improve the model's robustness to external attacks.	Siva Kumar et al. (2020), Anthi et al. (2021)
5.	Fault diagnosis might not be effective if the decision is made on monitoring a single sensor value.	Multi-sensor data fusion	Multi-sensor fusion involves tracking the machinery operation using varied sensors and making model predictions on the fused data from these sensors.	T. Li et al. (2020), Shan et al. (2020)

REFERENCES

Anthi, Eirini, Lowri Williams, Matilda Rhode, Pete Burnap, and Adam Wedgbury. 2021. "Adversarial Attacks on Machine Learning Cybersecurity Defences in Industrial Control Systems". *Journal of Information Security and Applications* 58 (May). Elsevier Ltd: 102717. doi:10.1016/j.jisa.2020.102717.

Azamfar, Moslem, Jaskaran Singh, Inaki Bravo-Imaz, and Jay Lee. 2020. "Multisensor Data Fusion for Gearbox Fault Diagnosis Using 2-D Convolutional Neural Network and Motor Current Signature Analysis". *Mechanical Systems and Signal Processing* 144 (October). Academic Press: 106861. doi:10.1016/j.ymssp.2020.106861.

"Case Western Reserve University Bearing Data Center Website | Bearing Data Center". 2021. Accessed 29 April. https://csegroups.case.edu/bearingdatacenter/pages/welcome-case-western-reserve-university-bearing-data-center-website.

Chatterjee, Joyjit, and Nina Dethlefs. 2020. "Temporal Causal Inference in Wind Turbine SCADA Data Using Deep Learning for Explainable AI". *Journal of Physics: Conference Series* 1618. IOP Publishing Ltd: 22022. doi:10.1088/1742-6596/1618/2/022022.

Chen, Jiayu, Cuiying Lin, DI Peng, and Hongjuan Ge. 2020. "Fault Diagnosis of Rotating Machinery: A Review and Bibliometric Analysis". *IEEE Access* 8: 224985–3. doi:10.1109/ACCESS.2020.3043743.

Chen, Xuefeng, Shibin Wang, Baijie Qiao, and Qiang Chen. 2018. "Basic Research on Machinery Fault Diagnostics: Past, Present, and Future Trends". *Frontiers of Mechanical Engineering*. Higher Education Press. doi:10.1007/s11465-018-0472-3.

Dhanraj, Joshuva Arockia, V Sugumaran, and A Joshuva. 2021. "A Lazy Learning Approach for Condition Monitoring of Wind Turbine Blade Using Vibration Signals and Histogram Features Fault Diagnosis of Wind Turbine View Project Fault Diagnosis and Localization on Wind Turbine Blade View Project A Lazy Learning Approach for Condition Monitoring of Wind Turbine Blade Using Vibration Signals and Histogram Features". Accessed 3 May. doi:10.1016/j.measurement.2019.107295.

Ellefsen, Andre Listou, Vilmar Æsøy, Sergey Ushakov, and Houxiang Zhang. 2019. "A Comprehensive Survey of Prognostics and Health Management Based on Deep Learning for Autonomous Ships". *IEEE Transactions on Reliability* 68 (2): 720–40. doi:10.1109/TR.2019.2907402.

Eren, Levent, Turker Ince, and Serkan Kiranyaz. 2019. "A Generic Intelligent Bearing Fault Diagnosis System Using Compact Adaptive 1D CNN Classifier". *Journal of Signal Processing Systems* 91 (2). Springer New York LLC: 179–89. doi:10.1007/s11265-018-1378-3.

Janiesch, Christian, Patrick Zschech, and Kai Heinrich. 2021. "Machine Learning and Deep Learning". *Electronic Markets*, April. Springer Science and Business Media LLC, 1–11. doi:10.1007/s12525-021-00475-2.

Jasiulewicz-Kaczmarek, Małgorzata, and Arkadiusz Gola. 2019. "Maintenance 4.0 Technologies for Sustainable Manufacturing: An Overview". *IFAC-PapersOnLine* 52. Elsevier B.V: 91–96. doi:10.1016/j.ifacol.2019.10.005.

Jiang, Wenqian, Yang Hong, Beitong Zhou, Xin He, and Cheng Cheng. 2019. "A GAN-Based Anomaly Detection Approach for Imbalanced Industrial Time Series". *IEEE Access* 7. Institute of Electrical and Electronics Engineers Inc.: 143608–19. doi:10.1109/ACCESS.2019.2944689.

Kamat, Pooja and Rekha Sugandhi. 2020a. "Anomaly Detection for Predictive Maintenance in Industry 4.0: A Survey". In E3S Web of Conferences. EVF 2019 Pune. doi:10.1051/e3sconf/202017002007.

Kamat, Pooja and Rekha Sugandhi. 2020b. "DigitalCommons @ University of Nebraska – Lincoln Bibliometric Analysis of Bearing Fault Detection Using Artificial Intelligence". *Library Philosophy and Practice* 4350.

Krishnamurthy, Vikram, Kusha Nezafati, Erik Stayton, and Vikrant Singh. 2020. "Explainable AI Framework for Imaging-Based Predictive Maintenance for Automotive Applications and Beyond". *Data-Enabled Discovery and Applications* 4 (1). Springer Science and Business Media LLC: 1–15. doi:10.1007/s41688-020-00042-2.

Li, Tianfu, Zhibin Zhao, Chuang Sun, Ruqiang Yan, and Xuefeng Chen. 2020. "Multi-Scale CNN for Multi-Sensor Feature Fusion in Helical Gear Fault Detection". *Procedia Manufacturing* 49. Elsevier B.V.: 89–93. doi:10.1016/j.promfg.2020.07.001.

Li, Xiang, Wei Zhang, Qian Ding, and Jian Qiao Sun. 2019. "Multi-Layer Domain Adaptation Method for Rolling Bearing Fault Diagnosis". *Signal Processing* 157 (April). Elsevier B.V.: 180–97. doi:10.1016/j.sigpro.2018.12.005.

Mao, Wentao, Yamin Liu, Ling Ding, and Yuan Li. 2019. "Imbalanced Fault Diagnosis of Rolling Bearing Based on Generative Adversarial Network: A Comparative Study". *IEEE Access* 7. Institute of Electrical and Electronics Engineers Inc.: 9515–30. doi:10.1109/ACCESS.2018.2890693.

Mao, Wentao, Bin Sun, and Liyun Wang. 2021. "A New Deep Dual Temporal Domain Adaptation Method for Online Detection of Bearings Early Fault". *Entropy* 23 (2). MDPI AG: 162. doi:10.3390/e23020162.

Ou, Jiayu, Hongkun Li, Gangjin Huang, and Guowei Yang. 2021. "Intelligent Analysis of Tool Wear State Using Stacked Denoising Autoencoder with Online Sequential-Extreme Learning Machine". *Measurement* 167: 108153. doi:10.1016/j.measurement.2020.108153.

Pimentel, Marco A F, David A Clifton, Lei Clifton, and Lionel Tarassenko. 2014. "A Review of Novelty Detection". *Signal Processing* 99: 215–49. doi:10.1016/j.sigpro.2013.12.026.

Sartal, Antonio, Roberto Bellas, Ana M Mejías, and Alberto García-Collado. 2020. "The Sustainable Manufacturing Concept, Evolution and Opportunities within Industry 4.0: A Literature Review". *Advances in Mechanical Engineering* 12 (5). SAGE Publications Inc.: 168781402092523. doi:10.1177/1687814020925232.

Saufi, Syahril Ramadhan, Zair Asrar Bin Ahmad, Mohd Salman Leong, and Meng Hee Lim. 2019. "Challenges and Opportunities of Deep Learning Models for Machinery Fault Detection and Diagnosis: A Review". *IEEE Access* 7: 122644–62. doi:10.1109/ACCESS.2019.2938227.

Shan, Pengfei, Hui Lv, Linming Yu, Honghong Ge, Yang Li, and Le Gu. 2020. "A Multisensor Data Fusion Method for Ball Screw Fault Diagnosis Based on Convolutional Neural Network with Selected Channels". *IEEE Sensors Journal* 20 (14). Institute of Electrical and Electronics Engineers Inc.: 7896–7905. doi:10.1109/JSEN.2020.2980868.

Shi, Huaitao, Jin Guo, Zhe Yuan, Zhenpeng Liu, Maxiao Hou, and Jie Sun. 2020. "Incipient Fault Detection of Rolling Element Bearings Based on Deep EMD-PCA Algorithm". *Shock and Vibration* 2020: 8871433. doi:10.1155/2020/8871433.

Siva Kumar, Ram Shankar, Magnus Nystrom, John Lambert, Andrew Marshall, Mario Goertzel, Andi Comissoneru, Matt Swann, and Sharon Xia. 2020. "Adversarial Machine Learning: Industry Perspectives". In *Proceedings – 2020 IEEE Symposium on Security and Privacy Workshops, SPW 2020*, 69–75. Institute of Electrical and Electronics Engineers Inc. doi:10.1109/SPW50608.2020.00028.

Susto, Gian Antonio, Andrea Schirru, Simone Pampuri, Seán McLoone, and Alessandro Beghi. 2015. "Machine Learning for Predictive Maintenance: A Multiple Classifier Approach". *IEEE Transactions on Industrial Informatics* 11 (3). IEEE Computer Society: 812–20. doi:10.1109/TII.2014.2349359.

Tao, Xinmin, Chao Ren, Yongkang Wu, Qing Li, Wenjie Guo, Rui Liu, Qing He, and Junrong Zou. 2020. "Bearings Fault Detection Using Wavelet Transform and Generalized Gaussian Density Modeling". *Measurement: Journal of the International Measurement Confederation* 155. Elsevier Ltd: 107557. doi:10.1016/j.measurement.2020.107557.

Tian, Jing, Lili Liu, Fengling Zhang, Yanting Ai, Rui Wang, and Chengwei Fei. 2020. "Multi-Domain Entropy-Random Forest Method for the Fusion Diagnosis of Inter-Shaft Bearing Faults with Acoustic Emission Signals". *Entropy* 22 (1): 57. doi:10.3390/e22010057.

Wang, Huaqing, Shi Li, Liuyang Song, and Lingli Cui. 2019. "A Novel Convolutional Neural Network Based Fault Recognition Method via Image Fusion of Multi-Vibration-Signals". *Computers in Industry* 105 (February). Elsevier B.V.: 182–90. doi:10.1016/j.compind.2018.12.013.

Wang, Wenjian, Zongben Xu, Weizhen Lu, and Xiaoyun Zhang. 2003. "Determination of the Spread Parameter in the Gaussian Kernel for Classification and Regression". *Neurocomputing* 55 (3–4). Elsevier: 643–63. doi:10.1016/S0925-2312(02)00632-X.

Wollenhaupt, Gary. 2016. "IoT Slashes Downtime with Predictive Maintenance". http://www.ptc.com/product-lifecycle-report/iot-slashes-downtime-with-predictive-maintenance.

Wuest, Thorsten, Daniel Weimer, Christopher Irgens, and Klaus Dieter Thoben. 2016. "Machine Learning in Manufacturing: Advantages, Challenges, and Applications". *Production and Manufacturing Research* 4 (1): 23–45. doi:10.1080/21693277.2016.1192517.

Yu, Wenjin, Tharam Dillon, Fahed Mostafa, Wenny Rahayu, and Yuehua Liu. 2020. "A Global Manufacturing Big Data Ecosystem for Fault Detection in Predictive Maintenance". *IEEE Transactions on Industrial Informatics* 16 (1). IEEE Computer Society: 183–92. doi:10.1109/TII.2019.2915846.

Yuan, Laohu, Dongshan Lian, Xue Kang, Yuanqiang Chen, and Kejia Zhai. 2020. "Rolling Bearing Fault Diagnosis Based on Convolutional Neural Network and Support Vector Machine". *IEEE Access* 8: 137395–406. doi:10.1109/ACCESS.2020.3012053.

Zebari, Rizgar, Adnan Abdulazeez, Diyar Zeebaree, Dilovan Zebari, and Jwan Saeed. 2020. "A Comprehensive Review of Dimensionality Reduction Techniques for Feature Selection and Feature Extraction". *Journal of Applied Science and Technology Trends* 1 (2): 56–70. doi:10.38094/jastt1224.

Zhang, Yongxiang, and R. B. Randall. 2009. "Rolling Element Bearing Fault Diagnosis Based on the Combination of Genetic Algorithms and Fast Kurtogram". *Mechanical Systems and Signal Processing* 23 (5): 1509–17. doi:10.1016/j.ymssp.2009.02.003.

Zonta, Tiago, Cristiano André da Costa, Rodrigo da Rosa Righi, Miromar José de Lima, Eduardo Silveira da Trindade, and Guann Pyng Li. 2020. "Predictive Maintenance in the Industry 4.0: A Systematic Literature Review". *Computers and Industrial Engineering* 150 (December). Elsevier Ltd: 106889. doi:10.1016/j.cie.2020.106889.

5 A Multi-Agent Reinforcement Learning Approach for Spatiotemporal Sensing Application in Precision Agriculture

T. A. Tamba

CONTENTS

5.1 INTRODUCTION: BACKGROUND AND DRIVING FORCES

Digital transformations within the realm of Industry 4.0 have introduced a paradigm shift in the management and production systems of small and medium-sized enterprises (SMEs) in the agricultural sector. Embracing such ideas as "smart farming" and "precision agriculture", farmers and agricultural SMEs are now using wireless sensor networks (WSN) and autonomous robots (e.g. wheeled mobile robots and drones) to help maximise their production output while at the same time minimising farming costs (Liakos et al. 2018). One characteristic of such networks is that they are often implemented as a wireless sensor network system whereby the individual

DOI: 10.1201/9781003200857-5

sensor (or mobile robot) in the network has only limited sensing or movement coverage abilities. In this regard, an important challenge in the implementation of such WSNs is the so-called optimal coverage problem which is essentially concerned with addressing the question of how to coordinate each individual sensor (or mobile robot) that has partial sensing (or movement coverage) ability with other sensors to ensure that their resulting network is capable of providing full area sensing (or movement) coverage of a farming field (Pham et al. 2020, Mekonnen et al. 2019).

The area coverage problem frequently arises in various applications that involve the use of sensor networks to perform the monitoring and surveillance of spatiotemporally varying variables/parameters. In general, each sensor in such networks has limited capacities (in terms of power, sensing distance, communication range, etc.) and therefore should be deployed in groups either in a static (installed at a fixed position/location) or in a dynamic (moves along a certain trajectory) manner (Barrientos et al. 2011). Regardless of the chosen deployment scheme, it is always important to ensure that the collective sensing of such statically/dynamically placed sensors can cover the desired area to be monitored.

One of the main challenges in addressing the area coverage problem of a WSN system in smart farming applications is the abundance of available sensor data as well as the high variability in the environmental conditions where the WSNs are deployed. In this regard, data-driven methods based on machine learning techniques have emerged as promising tools and approaches to addressing the various issues regarding the implementation of WSNs in smart farming, including that of the area coverage problem (Galceran et al. 2013, Asadi 2020, Lin et al. 2020).

This chapter aims to describe and review a machine learning approach called the multi-agent reinforcement learning (MARL) method for addressing the area coverage problem that often arises in the implementation of WSNs for smart farming applications (Nowé et al. 2012, van Otterlo et al. 2012, Greenwald et al. 2003, Littman 1994, 2001). More specifically, this chapter considers the case when a group of unmanned aerial vehicles (UAVs) is used as a (moving) WSN that performs the monitoring of a particular farm land. To formulate such a coverage problem in the MARL-based formalism, the UAVs are viewed as agents that interact with their environment (i.e. the farm land). The goal of such UAVs is determined to be that of maximising the overall coverage area monitored by the UAVs (formulated in the form of the field of view [FOV] of a camera attached to each UAV) while at the same time minimising possible overlaps of different UAVs in the group (Pham et al. 2018, 2020). The task of each agent is thus to search for certain action policies through the Q-learning iterative process in a game theoretic framework that will ensure the attainment of the group's goal/objective (Nowé et al. 2012, van Otterlo et al. 2012). Numerical simulation results are presented to illustrate the proposed MARL-based area coverage problem.

5.2 SYSTEM MODEL

This section describes the modelling formalism that is used in this chapter, which includes the Markov decision process (MDP) and Markov games (MG).

5.2.1 Markov Decision Process and Reinforcement Learning

Consider an agent that is involved in a decision-making process over a discrete time sequence $t = \{1,2,\ldots, T\}$ according to an MDP defined below (Nowé et al. 2012, van Otterlo et al. 2012).

Definition 1 (MDP): *An MDP is a tuple* $M = S, s^1, A, R, T$ *in which* $S = \left\{s^1, s^2, \ldots, s^{n_s}\right\}$ *denotes a (finite) set of possible states;* $s^1 \in S$ *is the initial state;* $A = \left\{a^1, a^2, \ldots, a^{m_a}\right\}$ *is a (finite) set of possible actions;* $R: S \times A \rightarrow \mathbb{R}$ *denotes the reward function that maps each pair of state-action into a real-valued number; and* $T: S \times A \times S \rightarrow [0,1]$ *denotes the transition function which maps the state transition from* $s \in S$ *to* $s' \in S$ *under action* $a \in A$ *into a number between* 0 *and* 1.

According to Definition 1, the dynamics of the agent starts at the initial state $s_1 = s^1$. At each time $t = \{1,2,\ldots, T\}$, the agent observes its current state $s_t \in S$, takes a particular action $a_t \in A$, and then receives feedback in the form of a reward $r_t \in R$. A state transition to $s_{t+1} = s' \in S$ is then observed and the agent repeats the decision-making process iteratively. Both the reward and transition functions in $\mathcal{M}$ satisfy the Markov property. Specifically, the expected reward at time t is dependent only on the present state-action pair and not on their history up to time t, i.e.

$$R(s,a) \cong \mathbb{E}\left[r_{t+1} \mid (s_t, s_{t-1}, \ldots, s_1), (a, a_{t-1}, \ldots, a_1)\right] = \mathbb{E}\left[r_{t+1} \mid s_t = s, a_{t=a}\right] \quad (5.1)$$

Similarly, the probability of transitioning at time t to a particular state s' is dependent only on the present state-action pair and not on their history up to time t, i.e.

$$\begin{aligned} T(s,a,s') &\cong \mathbb{P}\left[s_{t+1} = s' \middle| (s_t, s_{t-1}, \ldots, s_1), (a_t, a_{t-1}, \ldots, a_1)\right] \\ &= \mathbb{P}\left[s_{t+1} = s' \middle| s_t = s, a_t = a\right] \end{aligned} \quad (5.2)$$

For all $t = \{1,2,\ldots, T\}$, the choice of action taken by an agent at any time t is specified by the policy function $\pi{:}S \times A \rightarrow [0,1]$ which maps each element of the joint state-action into a number between 0 and 1. Thus, if at time t the agent is at the state $s_t = s$ and takes an action $a_t = \pi_t$, it can be said that the agent follows a policy $\pi_t(s)$.

In the MDP model, an agent's main objective is to find an optimal policy that will result in an optimal decision-making process. Such an optimal policy is usually computed using some reward-based optimisation method whereby the optimal policy π^* is defined as that which guarantees the maximisation of the expected total rewards $\nu^*(s)$ with a discounting factor γ defined as follows:

$$v^*(s) = v_{\pi^*}(s) \cong \max_{\pi} \mathbb{E}_{\pi}\left[\sum_{k=0}^{H} \gamma^k r_{t+k} \mid s_t = s\right] \quad (5.3)$$

in which π denotes the agent's policy; $\nu^*(s)$ denotes the obtained reward under the optimal policy π^*; $\mathbb{E}_{\pi}$ is the expected value operator with respect to policy π; k is a future time step of length/horizon H; and r_{t+k} denotes the reward within k time step in the future. The discount factor $\gamma \in [0,1]$ scales the significance of the reward k time step in the future.

One method to achieve the aforementioned objective of the agent in MDP is to use reinforcement learning (RL). In RL, the search for the agent's optimal policy is done through a learning process using as a reference the so-called optimal immediate Q-value $Q^*(s,a)$ of each pair of state-action $(s_t,a_t) = (s,a)$ at time t below (van Otterlo et al. 2012):

$$Q^*(s,a) = r_t(s,a) + \gamma \sum_{s'} T(s,a,s') \max_{a'} Q^*(s',a') \tag{5.4}$$

where $(s',a') = (s_{t+k},a_{t+k})$ is the state-action pair at each $k = 0,\ldots, H$ time step in the future; $r_t(s,a)$ is the agent's immediate reward at time t; and $T(s,a,s')$ is the agent's transition probability s to s' when choosing an action a. Since its exact value computation is challenging, an estimate $Q(s,a)$ of the optimal Q-value in Eq. (5.4) is often computed instead using, for instance, the well-known Q-learning iteration algorithm:

$$Q(s,a) \leftarrow (1-\alpha)Q(s,a) + \alpha\left[r(s,a) + \gamma \max_{a'} Q^*(s',a')\right] \tag{5.5}$$

in which $\alpha \in [0,1]$ is known as the *learning rate*. The convergence of the Q-value estimate in iteration (5.5) towards the optimal value defined in Eq. (5.4) was shown in Watkins and Dayan (1992) under the condition/assumption that the iteration visits all pairs of state-action infinitely often with an appropriate learning rate.

5.2.2 Multi-Agent Reinforcement Learning

The previously described RL scheme only describes the process of decision-making for one agent. When the decision-making process involves several agents simultaneously, a generalisation of the single-agent RL method within the framework of multi-agent reinforcement learning is needed.

Consider several interacting agents whose individual evolution is modelled as an MDP. One approach to describe the collective decision-making process of all such agents is by using the Markov game model. In essence, the MG model consists of multiple MDPs wherein the transition probability and reward of each agent now depend on the combined state-action pairs of all agents as defined in Definition 2 (Nowé et al. 2012, van Otterlo et al. 2012).

Definition 2 (MG): *An MG of $n \geq 2$ agents is a tuple $\mathcal{G} = n, \{S_i\}_{i=1}^n, \{s_i^1\}_{i=1}^n, \{A_i\}_{i=1}^n, \{R_i\}_{i=1}^n, \{T_i\}_{i=1}^n$ in which the set $\{S_i\}_{i=1}^n$ denotes all agents' state spaces such that $\mathbb{S} = \{S_i\}_{i=1}^n$ is the MG's state space; $\{s_i^1\}_{i=1}^n$ is the set of initial states of all agents; $\{A_i\}_{i=1}^n$ denotes all agents' set of actions such that $\mathbb{A} := \{A_i\}_{i=1}^n$ is the MG's action space; $R_i : \mathbb{S} \times \mathbb{A} \rightarrow \mathbb{R}$ denotes the i-th agent's reward function; and $T_i : \mathbb{S} \times \mathbb{A} \times \mathbb{S} \rightarrow [0,1]$ denotes the i-th agent's transition function.*

Definition 2 implies that the action choice of each agent depends on the combined pairs of state-action of all agents in $\mathcal{G}$. Consequently, the state transition of $\mathcal{G}$ is now built upon the actions of all agents and so in choosing the action at a particular state, each agent must take into consideration the action choices of other agents. This essentially means that each agent's choice of action at a certain state is determined as the solution of a repeated game whereby an agent acts as a player that chooses to coordinate/compete with other agents/players' joint state-action choices (Nowé et al. 2012, Littman 1994). In this regard, each agent's choice of action at each state in the MG evolution can be determined using methods from game theory to ensure that all agents arrive at an equilibrium joint policy.

For $i = 1,\ldots, n$, let $\pi_i : \mathbb{S} \times A_i \rightarrow \left[0,1\right]$ be the policy of each agent in $\mathcal{G}$ and $\pi_{\mathcal{G}} = \left(\pi_1,\ldots,\pi_n\right)$ be the joint actions of all such agents. The MARL framework defines the Q-value of the joint state-action pairs of agent i in $\mathcal{G}$ as follows:

$$Q_i^{\pi_{\mathcal{G}}}\left(\hat{s},\hat{a}\right) = \mathbb{E}_{\pi_{\mathcal{G}}}\left[\sum_{k=0}^{H} \gamma^k r_{t+k}^i \middle| \hat{s}_t = \hat{s}, \hat{a}_t = \hat{a}\right] \tag{5.6}$$

where $\hat{s}_t \in \mathbb{S}$ and $\hat{a}_t \in \mathbb{A}$ denote, respectively, the joint state and joint action of agent i at time t; and r_{t+k}^i denotes agent i's reward received k time steps in the future. It can be seen in Eq. (5.6) that the one-shot Q-value computation of each agent at a certain state is more involved since it must now be done over the extended joint spaces and actions of all agents in $\mathcal{G}$. Furthermore, in accordance with the game-theoretic solution concept, the objective of all agents in $\mathcal{G}$ is no longer to find an optimal joint policy but instead to find an equilibrium joint policy $\bar{\pi}_{\mathcal{G}}$. More specifically, such an equilibrium joint policy $\bar{\pi}_{\mathcal{G}}$ is defined as the solution of the one-shot game played by all agents at each joint state of $\mathcal{G}$.

Several concepts of equilibrium joint policy have been proposed in the literature, including among others the minimax-Q (Littman 1994, 2001), NashQ (Hu and Wellman 1998, 2003), FFQ and correlated-Q (Greenwald et al. 2003), asymmetric-Q (Könönen 2004), and adaptive optimal learning (Wang and Sandholm 2002) equilibrium concepts. As illustrated in Figure 5.1, the computation of the joint equilibrium of $\mathcal{G}$ in most of these works is formulated in a manner that is similar to the computation of the optimal policy in the single-agent RL scheme. In particular, the update of the Q-value of each individual agent i is defined as (cf. Step 4 in Figure 5.1 [Hu et al. 2015]))

$$Q_i\left(\hat{s},\hat{a}\right) \leftarrow \left(1-\alpha\right)Q_i\left(\hat{s},\hat{a}\right) + \alpha\left[r_i\left(\hat{s},\hat{a}\right) + \gamma\Phi_i(\hat{s}')\right] \tag{5.7}$$

where $\Phi_i\left(\hat{s}'\right)$ denotes the expectation of the equilibrium value in the joint states $\hat{s}' = \hat{s}_{t+k}$ at the next k time steps for agent i. In general, computation of the equilibrium of the joint states is done using the solution framework to find mixed strategy equilibrium in a multi-player game. Specifically, each agent/player computes its optimal action using information obtained from its generated replicate of other agents/players' reward functions. Some well-known issues in solving a mixed strategy game

Input: element of the MG $\mathcal{G}$: agent set $i = 1, \cdots, n$, $\mathbb{S}$ (joint state), $\mathbb{A}$ (joint action), α (learning rate), γ (discounting factor), greedy search coefficient ϵ, number of episodes, Ξ

Initialization: for all $s_i^1 \in \mathbb{S}$, $\hat{a}_t \in \mathbb{A}$ with $i = 1, \cdots, n$, set $Q_i(\hat{s}, \hat{a}) \leftarrow 0$

for-each $\kappa = 1, \cdots, \Xi$ **do**
  initialize the state set s_i^1
  repeat
    Step-1: $\mathcal{G}_t = \langle n, \mathbb{S}, \mathbb{A}, Q_i(\hat{s}_t, \hat{a}_t), T_i(\hat{s}_t, \hat{a}_t) \rangle$ *% set normal-form game*
    Step-2: $\Phi_i(\widehat{s'}) \leftarrow \mathcal{G}_t$ with ϵ-greedy policy *% find the mixed strategy equilibrium* $\Phi_i(\cdot)$
    Step-3: $\hat{a} \leftarrow \Phi_i(\widehat{s'})$ *% choose joint action based on* $\Phi_i(\cdot)$
    Step-4: $Q_i(\hat{s}, \hat{a}) \leftarrow (1-\alpha) Q_i(\hat{s}, \hat{a})$
      $+\alpha \left[r_i(\hat{s}, \hat{a}) + \gamma \Phi_i(\widehat{s'}) \right]$ *% update each agent's Q-value of*
    Step-5: $\hat{s} \leftarrow \widehat{s'}$ *% update the value of the next joint state*
  until $\hat{s} = \hat{s}_{t+H}$

FIGURE 5.1 Equilibrium-based joint policy computation algorithms in MARL.

such as a computationally intensive equilibrium search as well as reward function sharing requirements make the problem of solving MARL using a game theoretic framework open.

5.3 A MARL-BASED AREA COVERAGE METHOD

The MARL framework can be used to formulate and provide a solution method to address such a sensor area coverage problem. One approach to formulating the area coverage problem of a sensor network within the MARL scheme is by viewing the sensors in the network as a multi-agent system. In particular, each such sensor is required to achieve both the local (individual) and global (team) objectives. For instance, the local objective may require each sensor to cover a subset region of the target area beyond a certain minimum threshold value. For the global objective, each sensor may be required to coordinate with other sensors to ensure that the resulting network can cover the whole target area with minimum coverage overlaps while guaranteeing an acceptable level of inter-agent communication system quality.

For a given number of sensors that are placed at a particular initial position configuration, the area coverage problem is thus concerned with how these sensors should be rearranged or reconfigured to ensure that each sensor achieves both the local and the global objectives. It can be seen, however, that the Q-value iteration of the MARL scheme formulated in Eq. (5.7) only captures the local objective of each agent through the individual reward term $r_i(\hat{s}, \hat{a})$. To take the global (team) objective of all agents into account, an agent may instead replace Eq. (5.7) with a global (team) reward $\mathcal{R}(\hat{s}, \hat{a})$ such that the Q-value iteration of the game becomes

$$Q_i(\hat{s}, \hat{a}) \leftarrow (1-\alpha) Q_i(\hat{s}, \hat{a}) + \alpha \left[\mathcal{R}(\hat{s}, \hat{a}) + \gamma \Phi_i(\hat{s}') \right] \tag{5.8}$$

The global reward $\mathcal{R}(\hat{s},\hat{a})$ may, for instance, be defined as a weighted combination of the agent's local reward at each time instance k. As suggested in Sadhu and Konar (2017), such a use of the global reward function in the Q-value update can help the iteration converge faster, allowing the team to achieve the objective quicker. Chapter 6 presents an illustration of the MARL-based formulation of the area coverage problem using multiple UAVs as sensors.

5.3.1 Problem Description

Consider a planar target area $\mathcal{F}$ of arbitrary shape to be covered by a sensor network. To formulate the area coverage problem of $\mathcal{F}$ using the MARL framework, define a partition of $\mathcal{F}$ into a number of rectangular grid cells of equal area. Assume that a group of UAVs, each of which is attached with a camera pointing downward, is to be deployed to perform the coverage of $\mathcal{F}$, cf. Figure 5.2 (adapted from Pham et al. 2018). We assume the following characteristics of each UAV and the camera attached to it:

- Each UAV is equipped with both a localisation system that provides its three-dimensional Cartesian position and an onboard motion controller that controls its movements in three dimensions.

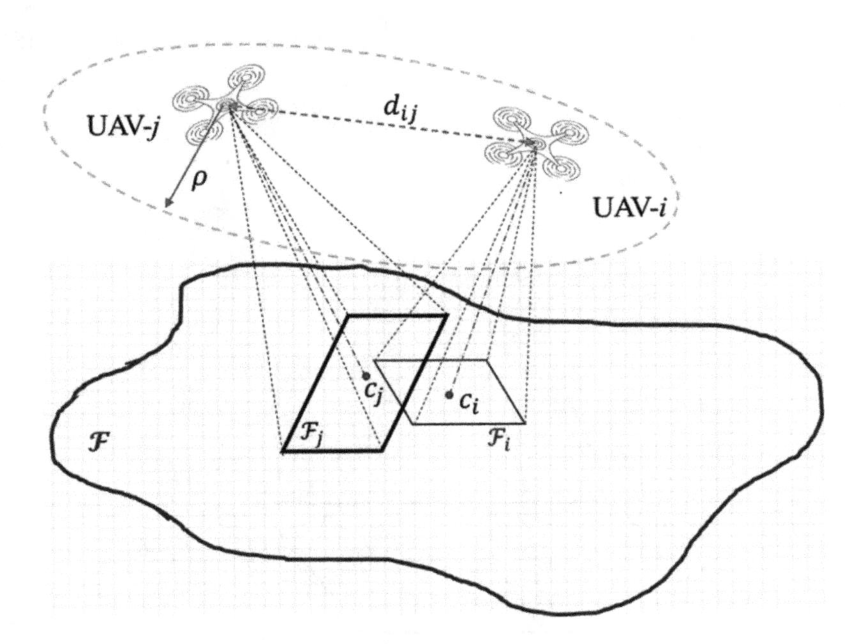

FIGURE 5.2 Area coverage of camera-equipped multi-UAVs (adapted from Pham et al. 2018).

- Each UAV has a sphere-shaped communication range of radius ρ within which communication and information transfer with other UAVs in the network are possible.
- The camera in each UAV provides a rectangular field of view with an area that can be enlarged (or reduced) by increasing (or decreasing) the altitude of the UAV.

Let there be n UAVs used in the network and denote with $p_i = (x_i, y_i, z_i)^T$ for $i = 1,\ldots, n$ the three-dimensional position of each UAV. For the camera of the i-th UAV, denote with $\mathcal{F}_i$ the rectangular FOV it covers with a centre c_i that is defined by the projection of the UAV's planar position $(x_i, y_i)^T$ onto the target area $\mathcal{F}$. The position p_i of the i-th UAV and the FOV $\mathcal{F}_i$ covered by the camera attached to it thus form a pyramid of height z_i and half-angles $\theta_i = \left(\theta_i^1, \theta_i^2\right)^T$ as illustrated in Figure 5.3 (adapted from Pham et al. 2018). Thus, a point $q \in \mathcal{F}$ in the target area is covered by the i-th UAV if Eq. (5.9) holds for each i.

Given two UAVs with a distance d_{ij} from one another, let $\mathcal{F}_{ij} = \mathcal{F}_i \cap \mathcal{F}_j$ for each pair of $(i,j) \in 1,\ldots, n$ with $i \neq j$ be the overlapping region of their FOVs. For the local objective, assume that the FOV of each UAV is required to cover a region that is larger than a threshold value Δ, that is $\mathcal{F}_{ij} \geq \Delta$ with $\Delta > 0$ a constant. Furthermore, let the global objective of the n UAVs be defined as follows (Pham et al. 2018):

$$C = \int_{q \in \mathcal{F}} \mathcal{A}(q, p_1, \ldots, p_n)\mathcal{E}(q)dq - \int_{q \in \mathcal{F}_{ij}} \mathcal{A}(q, p_1, \ldots, p_n)\mathcal{E}(q)dq \tag{5.9}$$

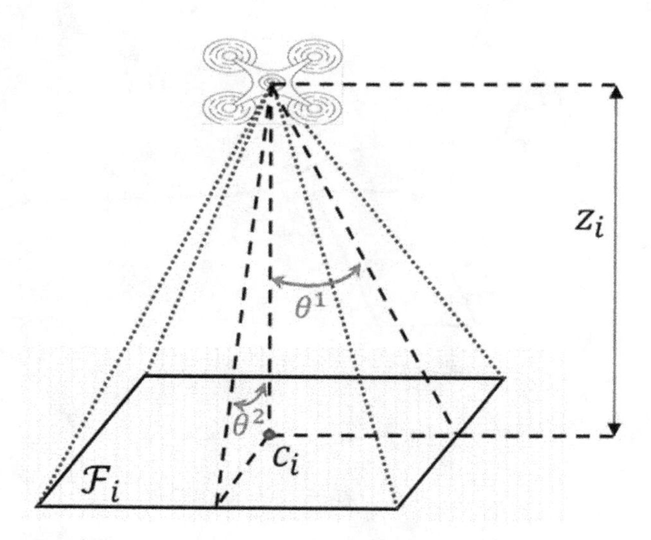

FIGURE 5.3 FOV of the camera on UAV I (adapted from Pham et al. 2018).

where $\mathcal{A}(q, p_1, \ldots, p_n)$ denotes the combined FOVs of the UAVs and $\varepsilon(q)$ is a measure of the importance of a particular region in $\mathcal{F}$. Note that the objective Eq. (5.9) captures the requirement that the group of such UAVs provides combined FOVs that fully cover the target area $\mathcal{F}$ with minimum overlaps. Given the description of the multi-UAVs system and the corresponding area coverage problem, the MARL framework described previously can be formulated and used to provide a solution approach as described in the following section.

5.4 A MARL-BASED AREA COVERAGE APPROACH WITH INTER-AGENT NEGOTIATION

We consider the problem of designing a decision-making strategy for a group of n UAVs that are tasked with covering an area $\mathcal{F}$ of arbitrary shape described in the previous section. More specifically, our objective is to examine the use of the MARL-based approach discussed in Section 5.2 to solve the area coverage problem formulated in Section 5.3.

5.4.1 Markov Game Model

Based on the area coverage problem in Section 5.3, we view the n UAVs as agents that operate on an environment that is defined by a three-dimensional box having a finite number of cubes. Each cube's position on the box is defined as the corresponding cube's centre point. The decision-making processes of the UAVs on the meshed box representation of the environment may thus be modelled as a Markov game: $\mathcal{G} = n, \{S_i\}_{i=1}^n, \{s_i^1\}_{i=1}^n, \{A_i\}_{i=1}^n, \{R_i\}_{i=1}^n, \{T_i\}_{i=1}^n$ where $i = 1, \ldots, n$ denotes the index of each UAV in the group. The set of possible states $\{S_i\}_{i=1}^n$ of each agent is determined as the three-dimensional position $p_i = (x_i, y_i, z_i)^T$ of each UAV in the box, where (x_i, y_i, z_i) is defined as the centre point of the cube where the UAV is located. The set of possible actions $\{A_i\}_{i=1}^n$ of each agent consists of six possible translational movements, namely heading in the east (HE), west (HW), north (HN), and south (HS) directions for lateral or horizontal movements, as well as up (U) and down (D) for altitude or vertical movements. It is set that, whenever an agent is on the cube (state) that is located at the boundary of the box and then chooses an action that takes it out of the box, then the agent should stay put and remain on that particular cube/state. Moreover, given the state set $\{S_i\}$ and the possible action set $\{A_i\}$, it is assumed that the transitions from a particular state $\hat{s}_t$ at time t to another state $\hat{s}_{t+1}$ under an action $\hat{a}_t$ have equal probability. The reward function of $\mathcal{G}$ is defined in connection with the group's coverage, objective function $\mathcal{C}$ in Eq. (5.9). On the one hand, for a given meshed box environment of UAVs, the group objective is to maximise the resulting coverage area with minimum overlaps, i.e.

$$\max_{\hat{s}_t \in \{S_i\}} \mathcal{C} = \underset{\hat{s}_t}{\operatorname{argmax}} \left\{ \sum_i^n \mathcal{A}_i(\hat{s}_t) - \sum_i \mathcal{O}_i(\hat{s}_t) \right\} \tag{5.10}$$

where $\mathcal{A}_i$ denotes the number of cubes that is covered by each UAV's FOV; and $\mathcal{O}_i$ denotes the number of cubes in the i-th UAV's FOV which overlaps with the cubes of other UAVs' FOVs. On the other hand, the group is set to receive a reward if their combined FOVs are greater than or equal to a predefined minimum coverage value $\mathcal{A}_{\min}$ with no overlaps, that is

$$\mathcal{R}(\hat{s},\hat{a}) = \begin{cases} R, & \text{if} \sum_i^n \mathcal{A}_i(\hat{s}_t) \geq \mathcal{A}_{\min}, \text{ and} \sum_i \mathcal{O}_i(\hat{s}_t) \leq 0, \\ 0, & \text{otherwise.} \end{cases} \tag{5.11}$$

5.4.2 Learning for Equilibrium Computation

Given the MG model as described in Section 5.4.1, the computation of the game's equilibrium can be performed using Q-value iterative learning in Eq. (5.8). As described before, one main challenge in performing such an iteration is the computational complexity in performing value function updates on each agent because the dimension of the considered joint state-action space grows in an exponential manner with respect to the increase in the number of agents involved in the game. To address such a potential issue, we propose the use of the so-called Negotiation-based Q-learning (NQ learning) approach proposed in Hu et al. (2015) and Zhou et al. (2016), which is realistic and suitable for a multi-robot operation problem as discussed in this chapter. In essence, the basic idea in NQ learning is to perform the value function updates only when necessary (under certain conditions) and with only a subset of other agents in the game. More specifically, the learning process of each agent is decomposed into two main stages as follows (Zhou et al. 2016):

- First, assuming the absence of other agents in the environment, each agent performs a static learning process to compute its own individual optimal policies.
- Second, each agent interacts with others in an MG setting and performs coordinated learning with some of the agents in the group in a game theoretic manner.

Note that the first stage is basically a single-agent Q-learning iteration that is performed by each agent in the group and can be done according to Eq. (5.5). Once the first stage is completed, each agent will have a record of rewards of its own greedy optimal policy that can be used as a starting point for deciding what to do during the interaction with other agents in the group. In the second stage, each agent acts in a multi-agent setting and continuously monitors the returned rewards. If the agent finds that the rewards in the multi-agent setting are the same as those obtained in the first stage, then the agent chooses to act independently based on the greedy optimal policy and reward model that is obtained in the first step. If, in the multi-agent setting, the agent instead detects any change in the returned rewards, then the agent chooses to coordinate with other agents in a game theoretic manner. Such coordination begins with sharing an agent's individual optimal state-action pair with other neighbouring agents, and receives the state-action pair of each member of those neighbouring agents (Hu et al. 2015, Zhou et al. 2016).

We point out that the chosen coordination is done only with neighbouring agents and not all agents in the group. Particularly in the multi-UAVs operation considered in this chapter, the neighbour of a UAV can be defined as other UAVs that are located within the communication radius of that particular UAV over which data transfer between UAVs via wireless communication system is possible. After all members of the neighbouring group have interchanged their state-action pairs, each agent extends the space of its own state-action pairs with those of other agents and then coordinates in a game theoretic manner to compute their equilibrium policies. Different types of equilibrium policies are available to be considered for such a game, including a non-strict equilibrium-dominating strategy profile (NS-EDSP) or a meta-equilibrium set (cf. Hu et al. 2015). Once equilibrium is achieved, each agent chooses an action and repeats a similar learning process (i.e. the sequence of reward change detection and multi-agent coordination) as it moves. Interested readers are referred to Zhou et al. (2016) and Hu et al. (2015) for a more detailed algorithm of the multi-agent coordinated learning process as discussed above.

5.4.3 A Numerical Simulation

This section briefly presents numerical simulation results to illustrate the implementation of the proposed MARL-based area coverage approach for a multi-UAVs operation. The simulation considered the operation of two UAVs with each FOV angle of $\theta_1 = \theta_2 = 30°$ to perform the area coverage task on a rectangular farming region with an area size of (50×20) m^2 over which UAVs can fly within 10 m high. Thus, the operating region is defined as a box with a length, width, and height of 50 m, 20 m, and 10 m, respectively. Our simulation set-up considered the meshed version of the box with a cube of size 1 m^3 for the MARL-based area coverage analysis.

In the individual agent learning process, a reward of 100 is given if it reaches its goal, a reward of −10 is given if it steps out of the box operating region, and a reward of −1 is given otherwise to account for the battery consumption of the UAV. For simplicity, we assume that both UAVs are flying at a similar constant altitude of 5 m high such that the resulting operating region can simply be considered as a planar two-dimensional region of the farming area. The group reward function is defined as in Eqs. (5.9) and (5.11). The parameters of the MARL Q-value iteration in Eq. (5.8) for equilibrium learning of the MG are as follows: $\alpha = 0.1$, $\gamma = 0.9$, greedy search coefficient $\epsilon = 0.01$, and number of learning episodes $\Xi = 20$.

Figures 5.4 and 5.5 show, respectively, the obtained average reward of each agent as well as the group reward for the multi-UAVs system. It can be seen that each agent produces a policy that achieves both the individual and group returned rewards. Furthermore, the average number of steps required for the learning processes of both agents for the given number of episodes is shown in Figure 5.6. In particular, Figure 5.6 shows the convergence of the required number of steps under the increase in the

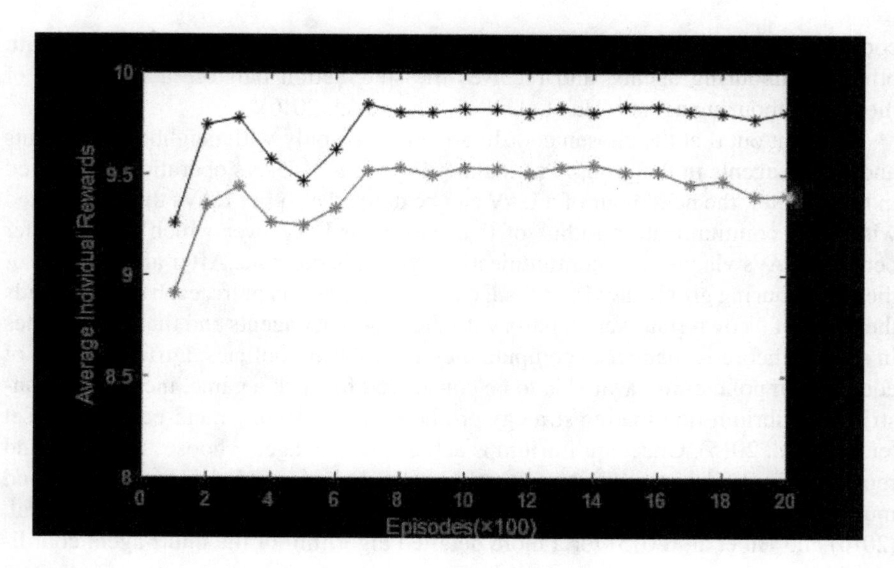

FIGURE 5.4 Average individual reward of each agent.

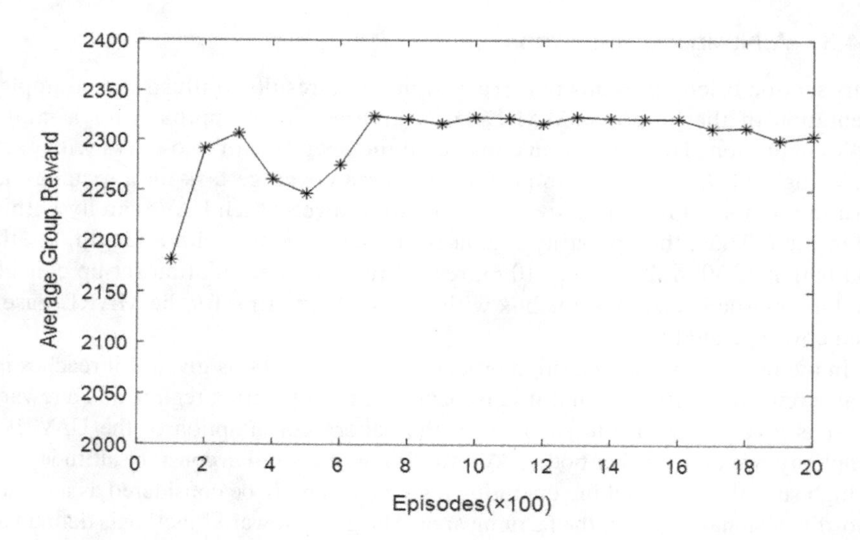

FIGURE 5.5 Average group reward of the multi-agent system.

chosen number of episodes. These results thus illustrate the effectiveness of the proposed multi-agent area coverage method for multi-UAVs operation.

5.5 CONCLUDING REMARKS

This chapter has presented a MARL scheme to address the area coverage problem in smart farming applications. The coverage problem in the proposed method is viewed

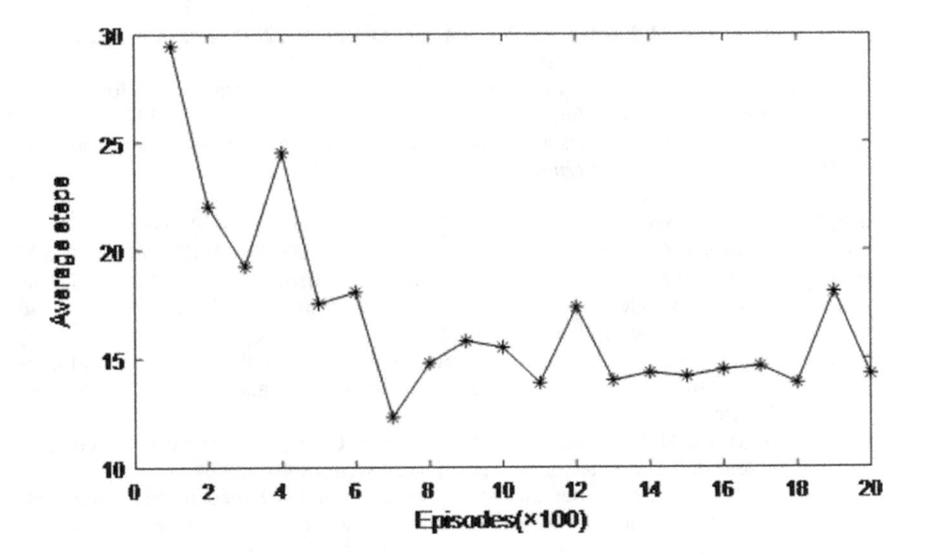

FIGURE 5.6 Average steps per episode.

as an interaction between autonomous agents (robots) and their environment (i.e. the farm land). The proposed method presents a scheme to search for action policies through a Q-learning iterative process in a game theoretic framework that ensures the attainment of maximal area coverage by all agents. Numerical simulation results are presented and illustrate the effectiveness of the proposed multi-agent area coverage method for multi-UAVs operation

REFERENCES

Asadi, R. 2020. Deep learning models for spatio-temporal forecasting and analysis. PhD diss., Univ. California, Irvine.

Barrientos, A. et al. 2011. Aerial remote sensing in agriculture: A practical approach to area coverage and path planning for fleets of mini aerial robots. *Journal of Field Robotics* 28, no. 5: 667–689.

Galceran, E. and M. Carreras. 2013. A survey on coverage path planning for robotics. *Robotics and Autonomous systems* 61, no. 12: 1258–1276.

Greenwald, A., K. Hall, and R. Serrano. 2003. Correlated Q-learning. *Proceedings of the 20th International Conference on Machine Learning* 3: 242–249, Washington DC, USA.

Hu, J. and M. P. Wellman. 1998. Multiagent reinforcement learning: Theoretical framework and an algorithm. *Proceedings of the 15th International Conference on Machine Learning* 98: 242–250, Madison, USA.

Hu, J. and M. P. Wellman. 2003. Nash Q-learning for general-sum stochastic games. *Journal of Machine Learning Research* 4, no. 11: 1039–1069.

Hu, Y. Y. Gao, and B. An. 2015. Multiagent reinforcement learning with unshared value functions. *IEEE Transactions on Cybernetics* 45, no. 4: 647–662.

Könönen, V. 2004. Asymmetric multiagent reinforcement learning. *Web Intelligence and Agent Systems: An International Journal* 2, no. 2: 105–121.

Liakos, K. G., P. Busato, D. Moshou, S. Pearson, and D. Bochtis. 2018. Machine learning in agriculture: A review. *Sensors* 18, no. 8: 2674.
Lin, T. et al. 2020. DeepCropNet: a deep spatial-temporal learning framework for county-level corn yield estimation. *Environmental Research Letters* 15, no. 3: 034016.
Littman, M. L. 1994. Markov games as a framework for multi-agent reinforcement learning. *Proceedings of the 11th International Conference on Machine Learning* 1: 157–163, New Brunswick, USA.
Littman, M. L. 2001. Friend-or-foe Q-learning in general-sum games. *Proceedings of the 18th International Conference on Machine Learning* 1: 322–328, Williamstown, USA.
Mekonnen, Y., S. Namuduri, L. Burton, A. Sarwat, and S. Bhansali. 2019. Machine learning techniques in wireless sensor network based precision agriculture. *Journal of the Electrochemical Society* 167, no. 3: 037522.
Nowé, A., P. Vrancx, and Y. M. D. Hauwere. 2012. Game theory and multi-agent reinforcement learning. In *Reinforcement Learning*, ed. M. Wiering and M. van Otterlo, 441–470. Berlin/Heidelberg: Springer.
Pham, H. X., H. M. La, D. F. Seifer, and A. Nefian. 2018. Cooperative and distributed reinforcement learning of drones for field coverage. ArXiv:1803.07250.
Pham, H. X., H. M. La, D. Feil-Seifer, and M. Deans. 2020. A distributed control framework of multiple unmanned aerial vehicles for dynamic wildfire tracking. *IEEE Transactions on Systems, Man, and Cybernetics: Systems* 50, no. 4: 1537–1548.
Sadhu, A. K. and A. Konar. 2017. Improving the speed of convergence of multi-agent Q-learning for cooperative task-planning by a robot-team. *Robotics and Autonomous Systems* 92, no. 6: 66–80.
van Otterlo, M. and M. Wiering. 2012. *Reinforcement Learning and Markov Decision Processes*. Berlin/Heidelberg: Springer.
Wang, X. and T. Sandholm. 2002. Reinforcement learning to play an optimal Nash equilibrium in team Markov games. *Advances in Neural Information Processing Systems* 15: 1603–1610.
Watkins, C. J. C. H. and P. Dayan. 1992. Q-learning. *Machine Learning* 8, no. 3–4: 279–292.
Zhou, L., P. Yang, C. Chen, and Y Gao. 2016. Multi-agent reinforcement learning with sparse interactions by negotiation and knowledge transfer. *IEEE Transactions on Cybernetics* 47, no. 5: 1238–1250.

6 Digital Twin of a Laboratory Gas Turbine Engine Using Deep Learning Framework

Richa Singh, P. S. V. Nataraj, and Arnab Maity

CONTENTS

DOI: 10.1201/9781003200857-6

6.1 INTRODUCTION

Gas turbine engines (GTEs) are efficient means of power production in transportation systems, chemical plants, aircraft industries, and marine engineering [1]. GTEs exhibit a complex structure and high non-linearity throughout their overall performance. GTEs operate over a wide flight envelope involving temperature, pressure, and load conditions at their extreme limits. These severe operating conditions pose a challenge in developing controllers that attain superior performance while retaining the system's reliable functioning and stability at a minimal overall cost. In the aviation maintenance industry, the cost of engine maintenance alone accounts for a significant portion of operational expenditures. The GTE can be shut down before a total breakdown develops if the monitoring system detects a malfunction early enough. Thus, the accurate inference of GTEs is significant in engineering disciplines that use computational models such as a digital twin to analyse and predict complex physical behaviour.

A digital twin is a virtual representation of a dynamical system, providing a significant way to achieve intelligent monitoring of an existing system. Digital twins use either a data-driven model or a mathematical model that is a sufficiently accurate depiction of the physical systems. A digital twin allows real-time monitoring of systems and processes by reducing/preventing unnecessary downtimes due to false alarms. It also enables timely data analysis to detect problems before they arise by scheduling preventive maintenance. Digital twin technology reduces system verification and testing costs while providing early insights into the system dynamics. The strategic trend of constructing digital twins and validating them with real-time data allows flights to remain airborne for less money. This adds significant value to airplanes by increasing their performance and reliability [2].

6.1.1 LITERATURE

Over the years, considerable effort has been made to understand engine dynamics and their representation with complex models. High-fidelity engine models and simulators that accurately describe the operating conditions in steady-state and transient operations are necessary for off-design performance prediction, performance deterioration, and control strategies. Modelling a system can be approached in two ways: (i) the physics-based approach and (ii) the empirical/data-driven approach. A physics-based model, also known as the first principle or white-box model, is utilised when detailed information about the physics of the system is available. In this method, the mathematical equations based on the conservation equation of continuity, momentum, and energy are used to derive the model dynamics, and underlying assumptions are made based on ideal conditions to deal with the complexities. Several works of literature introduce the physics-based modelling of GTE using the first principle method [3–5], computational fluid dynamics [6], and an analytical model [7]. However, developing an accurate mathematical model of a complex engine is close to impossible because of the non-linear nature of rotating components and their inherent dependencies; therefore, developing a digital twin with model-based approaches is a challenging task. The empirical/data-driven model, also known as the black-box model, utilises the experimental data to empirically set up a statistical correlation between input and output variables to describe the plant behaviour. Popular data-driven modelling approaches are system identification [8] and neural network [1].

Deep learning (DL) has encountered an enormous research resurgence and has been shown to deliver state-of-the-art results in a variety of industrial applications. Deep neural networks (DNNs) have gained interest due to their indispensable factors, namely (i) the availability of enormous data and (ii) their immense computing power. Recent advances in deep learning architectures have already made significant contributions to artificial intelligence in various disciplines. The usage of neural networks in various renewable energy applications was reported in a review [9]. Small heavy-duty industrial gas turbine dynamics were identified using a non-linear autoregressive model [10]. An artificial neural network (ANN) model of a single-shaft gas turbine was examined as a computational time, accuracy, and resilience alternative to physical models [11,12]. Deep networks were later studied, constructed on higher-order, and optimised to a greater extent with advancements in computing capacity and enormous data collection. Multi-step motion prediction modelling for dynamic systems over long horizons using deep learning is addressed in Refs. [13,14]. Recently, a report on various neural networks has been analysed to forecast waste heat recovery [15]. A digital twin–assisted fault diagnosis scheme is proposed in Ref. [16]. According to substantial research on DNNs, the profound learning framework has great potential to be considered a legitimate alternative to standard modelling and control methodologies, especially when there is minimal knowledge of system dynamics. To the author's knowledge, none of the existing literature addressed the application of the time-series long short-term memory (LSTM) network-based digital twin of a complex multi-input multi-output (MIMO) laboratory GTE.

6.1.2 Contributions

This chapter presents a data-driven digital twin of the laboratory GTE using an LSTM neural network. Developing a digital twin starts with the collection of large-scale data sets covering the entire operating range of the system by performing experimental runs. It consists of three inputs (i.e. fuel flow, ambient temperature, and inlet pressure) and seven outputs (i.e. shaft speed, temperature, and pressure parameters at each station of a GTE). The input parameters are mapped with output parameters via a deep neural network to capture the complex non-linear dynamics of GTE parameters. A DNN is configured with deep layers of LSTM cells and trained by selecting hyper-parameters to minimise the error between experimental and predicted output measurements over a large data set. Several experimental runs acquire this data set with different input parameters covering the complete spectrum of operations. The trained model is first tested on a new data set of the laboratory GTE and is ready for deployment in a real-time set-up. The performance of the LSTM-based digital twin has been compared against the non-linear autoregression with exogenous inputs (NARX) network-based model as a benchmark. However, the literature and simulation studies have found that a single NARX network is not efficient in predicting the seven GTE parameters simultaneously. Thus, seven different NARX networks have been developed for predicting the response. The root mean squared error (RMSE) is used as a prediction performance evaluation metric. The LSTM-based digital twin is further validated against a mathematical model of the laboratory GTE. The performance evaluation shows that the LSTM network–based digital twin outperforms in predicting the dynamic response in real time. The LSTM network–based digital twin of the laboratory gas turbine is found efficient in predicting the dynamic behaviour of various patterns of fuel flow to the engine. The developed digital twin has applications in sensor validation, system identification, engine health monitoring, fault detection and diagnosis, and design optimisation.

The novelty of the proposed work is that the LSTM-based digital twin of the laboratory GTE efficiently predicts the behaviour of the engine parameters by mapping only three inputs with seven output parameters for the entire operating range in real time. The user is not required to have an in-depth knowledge of the complete engine dynamics. Moreover, the proposed LSTM-based digital twin is lucrative because the process of model development (i) requires comparatively less time, unlike the first principles approach, and (ii) offers an efficient prediction of engine parameters over the entire operating range, unlike the conventional data-driven models.

The chapter is organised as follows: A brief explanation of DNNs and proposed frameworks are given in Section 6.2. The experimental set-up and its working principle are described in Section 6.3. The modelling approach using an LSTM deep neural network is explained in Section 6.4. The validation of the proposed digital twin is discussed in Section 6.5. Section 6.6 concludes the work and gives the future scope.

6.2 DEEP NEURAL NETWORK

Deep learning is a machine learning paradigm that encourages neural networks consisting of multiple hidden layers to learn and extract meaningful information from a large set of data with various levels of abstraction [17]. Amidst the empirical success of deep learning, the theoretical foundation is still in its adolescence. This section gives a brief mathematical analysis of an ANN and a DNN. The layers of DNN are made from nodes/neurons, as shown in Figure 6.1a. Basically, a neuron with k inputs transforming a set of $\boldsymbol{x} \in \mathbb{R}^k$, the input signal is a function:

$$\hat{\boldsymbol{x}} = f\left(b + \sum_{i=1}^{n} \left(\langle \vec{\boldsymbol{w}}_i, \vec{\boldsymbol{x}}_i \rangle\right)\right) \tag{6.1}$$

where $\vec{\boldsymbol{w}}$ is a weight vector; b is bias; $f : \mathbb{R}^k \Rightarrow \mathbb{R}$ is a non-linear transformation known as an activation function of the neuron; and $\langle \, , \, \rangle$ is a real scalar product. The weights and biases are the learnable parameters of the deep learning network.

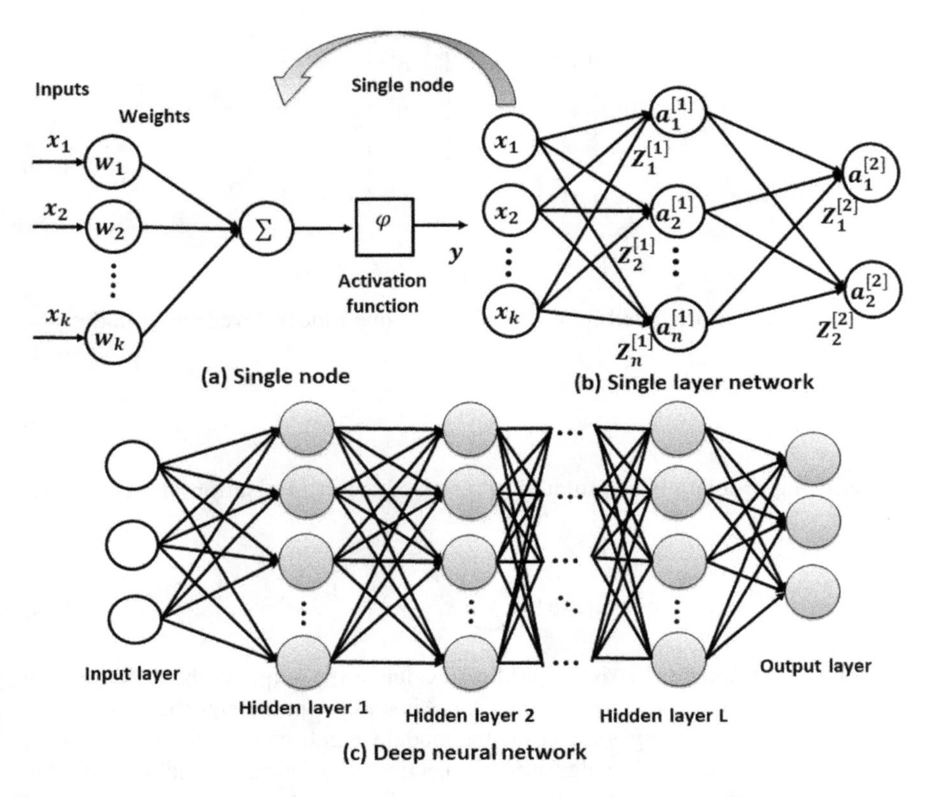

FIGURE 6.1 Schematic diagram for (a) one node of the neural network, (b) single-layer network, and (c) neural network.

The depth of a DNN distinguishes it from the more prevalent single hidden layer neural networks. DNNs aggregate and recombine features from previous layers, allowing the node to identify and learn more complex features from the data set as we advance deeper into the neural net [18]. Consider a shallow two-layer network, as shown in Figure 6.1b. The vector representation of a network with k as the input dimension, n nodes, and 2 as the output dimension is given for the first layer as

$$\begin{bmatrix} z_1^{[1]} \\ z_2^{[1]} \\ \vdots \\ z_n^{[1]} \end{bmatrix} = \begin{bmatrix} w_{11}^{[1]} & w_{12}^{[1]} \cdots & w_{1n}^{[1]} \\ w_{21}^{[1]} & w_{22}^{[1]} \cdots & w_{2n}^{[1]} \\ \vdots & \vdots \ \ddots & \vdots \\ w_{k1}^{[1]} & w_{k2}^{[1]} \cdots & w_{kn}^{[1]} \end{bmatrix}^T \begin{bmatrix} x_1 \\ x_2 \\ \vdots \\ x_k \end{bmatrix} + \begin{bmatrix} b_1^{[1]} \\ b_2^{[1]} \\ \vdots \\ b_n^{[1]} \end{bmatrix} \tag{6.2}$$

$$\begin{bmatrix} a_1^{[1]} & a_2^{[1]} \cdots & a_n^{[1]} \end{bmatrix}^T = f\begin{bmatrix} z_1^{[1]} & z_2^{[1]} \cdots & z_n^{[1]} \end{bmatrix}^T$$

and the second layer is given as

$$\begin{bmatrix} z_1^{[2]} \\ z_2^{[2]} \end{bmatrix} = \begin{bmatrix} w_{11}^{[2]} & w_{12}^{[2]} \\ w_{21}^{[2]} & w_{22}^{[2]} \\ \vdots & \vdots \\ w_{n1}^{[2]} & w_{n1}^{[2]} \end{bmatrix}^T \begin{bmatrix} a_1^{[1]} \\ a_2^{[1]} \\ \vdots \\ a_n^{[1]} \end{bmatrix} + \begin{bmatrix} b_1^{[2]} \\ b_2^{[2]} \\ \vdots \\ b_n^{[2]} \end{bmatrix} \tag{6.3}$$

$$\begin{bmatrix} a_1^{[2]} & a_2^{[2]} \end{bmatrix}^T = f\begin{bmatrix} z_1^{[2]} & z_2^{[2]} \end{bmatrix}^T$$

In the final output predicted by a network with one hidden layer and n nodes, the output layer is given by

$$\hat{y} = \begin{bmatrix} a_1^{[2]} \, a_2^{[2]} \end{bmatrix}^T \tag{6.4}$$

The learning problem is formulated in terms of the minimisation of a cost function, $\mathcal{L}$:

$$\min_{w,b} \mathcal{L}(w,b) = \sum_{i=1}^{n} (x_i - \hat{y}_i)^2 + \frac{1}{2}\lambda w^2 \tag{6.4}$$

where $\mathcal{L}$ is cost function; $\hat{y}_i$ is the predicted value of the output at the ith sample by the network; and λ is the learning rate. The λ is a hyper-parameter that determines how often the model changes each time the model weights change in response to the predicted error. The learning algorithm's objective is to determine the best possible values for the parameters w,b such that the overall loss of the network is minimised as much as possible.

The generalised formulation for a neural network with multiple hidden layers and multiple nodes, also known as a deep neural network in each of the layers, as shown in Figure 6.1c, is given as

$$a_n^L = \left[f\left(\sum_m w_{mn}^L \cdots \left[f\left(\sum_j w_{kj}^2 \left[f\left(\sum_k w_{ji}^1 x_i + b_j^1 \right) \right] + b_k^2 \right) \right] \cdots \right]_m + b_n^L \right) \Bigg]_n \tag{6.5}$$

where L is the number of layers in the network and n nodes in each layer.

6.2.1 Architecture of LSTM Network

In the realm of deep learning, LSTM is an extension of the recurrent neural network (RNN) architecture. It can process entire sequences of data as well as single data points. LSTM, a variant of RNN, comes with a new structure where operational gates define the weight [19]. An LSTM node/cell consists of a memory cell and three gate/channels viz. input, forget, and output gates. Input, forget, and output gates governs the flow of information through the memory cell when its output and forgotten.

The following equations describe how a memory cell is updated at each gate every time instant when passing through an LSTM cell. The following notations are used to explain the working of an LSTM cell:

1. x_t is the memory cell input at time t.
2. i_t, f_t, o_t, and C_t are outputs from the input gate (i), forget gate (f), output gate (o), and cell state (C), respectively.
3. C_t is the updated cell state.
4. W_i, W_f, W_C, and W_o are weight matrices.
5. b_i, b_f, b_c, and b_o are bias vectors.
6. h_{t-1} are values of previously hidden layers.

First, $h_{(t-1)}$ and x_t enter the forget gate and are activated through the sigmoid function to give an output value between 0 and 1 in the cell state $C_{(t-1)}$, i.e.

$$f_t = \sigma\left(W_f\left[h_{t-1}, x_t\right] + b_f\right) \tag{6.6}$$

The memory cell then decides what additional information is going to remain in the cell state. The input gate is driven by a sigmoid layer, and the *tanh* layer creates a temporary vector $\hat{C}_t$ that adds to the cell state as

$$\begin{aligned} i_t &= \sigma\left(W_i\left[h_{t-1}, x_t\right] + b_i\right) \\ \hat{C}_t &= tanh\left(W_c\left[h_{t-1}, x_t\right] + b_c\right) \end{aligned} \tag{6.7}$$

Now, the old cell state C_{t-1} is updated to C_t by combining outputs from forget gate f_t and input gate i_t as

$$C_t = i_t * \hat{C}_t + f_t * C_{t-1} \tag{6.8}$$

Lastly, the output will be based on the updated cell state and output gate where a part of the information is let out as

$$\begin{aligned} o_t &= \sigma\left(W_o\left[h_{t-1}, x_t\right] + b_o\right) \\ h_t &= o_t * tanh\left(C_t\right) \end{aligned} \tag{6.9}$$

where $\tan h(x) = (2/1 + e^{-2x})$ and $\sigma\left(x\right) = 1/1 + e^{-x}$ are the inherent activation function (Figure 6.2).

LSTM cells are trained via supervised learning on a succession of sequence-based training data sets, employing an optimisation technique such as gradient descent coupled with back-propagation across time to compute the gradients needed throughout the optimisation process.

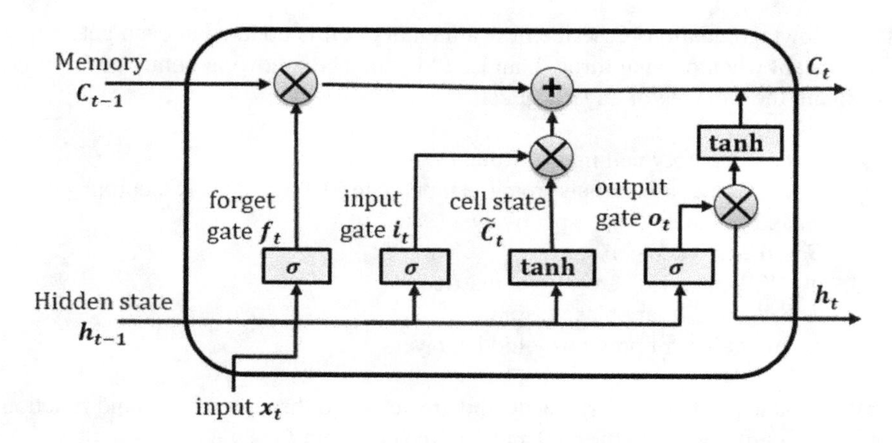

FIGURE 6.2 LSTM architecture with a memory cell, input gate, forget gate, and output gate.

6.3 SYSTEM DESCRIPTION AND PROBLEM FORMULATION

The laboratory GTE considered in this study is manufactured by Turbine Technologies, Ltd. [20] and is shown in Figure 6.3. A self-contained SR-30 jet engine and a TG-2000 electric generator are included in the Minilab configuration. The engine is installed with pressure transducers and thermocouples to keep track of the thermodynamic processes taking place within the engine. The SR-30 engine is an integrated unit consisting of an inlet nozzle, radial compressor, counterflow combustion chamber, axial flow turbine, and exhaust nozzle.

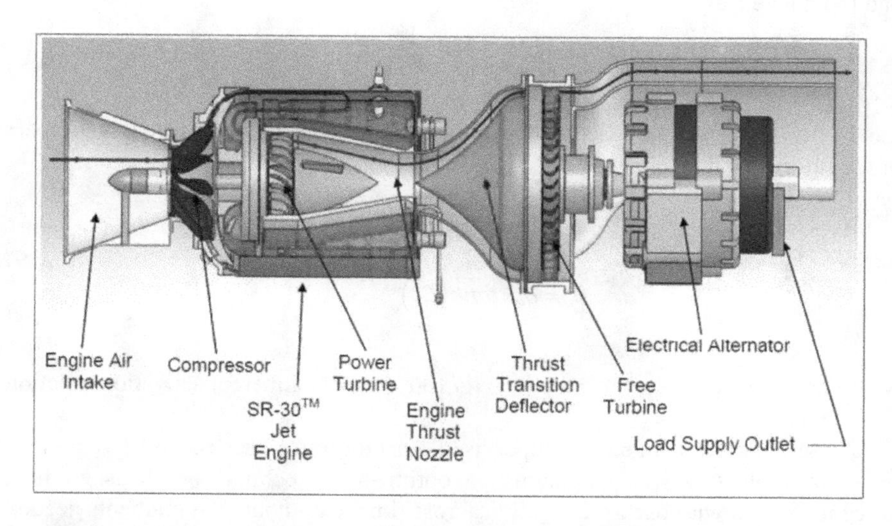

FIGURE 6.3 Cross-sectional view of laboratory SR-30 GTE [20].

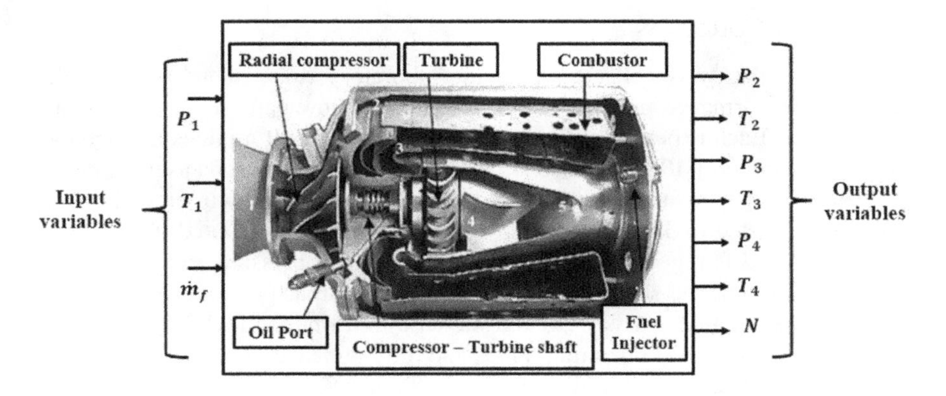

FIGURE 6.4 Input and output parameter of interest in SR-30 GTE.

The SR-30 laboratory engine's basic working is as follows: atmospheric air enters the compressor inlet and radially diverges in the centrifugal compressor where the kinetic energy is converted into a pressure head; the diffuser adjunct to the compressor increases the static pressure at the cost of kinetic energy. The air then mixes with fuel and ignites in the combustion chamber, and high-temperature gases expand in the turbine that produce work to drive the compressor, then accelerate through a nozzle producing thrust. The thermodynamic cycle is completed when the exhaust gas returns to ambient atmospheric conditions.

The Minilab configuration includes 13 sensors. The raw sensor package provides temperature, pressure, rpm, and flow rate sensor (calibrated) measuring parameters common to a Brayton Cycle analysis [6]. The DigiDAQ (data acquisition toolbox) system is compatible with LabVIEW from National Instruments. The sensors and actuators in the plant interface with the PC through National Instruments PXI Controller-8110, which is a 2.26 GHz Quad-Core Processor PXI Embedded Controller to enable bidirectional data flow.

In this chapter, the main objective is to utilise the pattern recognition ability over a time horizon of the LSTM network to predict engine parameters' dynamic response in real time. Given a series of fully observed time-series sensor measurements $x = x_1, x_2,\ldots, x_T$, where $x_t \in \mathbb{R}$, $t = 1,2,3,\ldots, T \in N$, and $N = 10$ is the input dimension, with which we aim at predicting a series of future signals. That being said, we developed a deep neural network with input and output, as shown in Figure 6.4, consisting of several neurons of LSTM cells in multiple hidden layers and trained it on the experimental data set with various hyper-parameters.

6.4 DIGITAL TWIN OF LABORATORY GTE

This section introduces the design process for the digital twin of a laboratory GTE using the LSTM network. The steps involved in developing an LSTM network–based digital twin of GTE using the aforementioned DL framework are discussed in the subsequent sections.

6.4.1 Data Collection

Experimental data was collected from the laboratory runs, intending to predict the engine performance parameters for various fuel flow patterns to the combustion chamber. Each experimental run lasts approximately 10 minutes and the data is logged at a rate of 100 milliseconds. Of the available sensor measurements, 10 parameters of interest are considered in developing the digital twin of a laboratory GTE and are listed in Table 6.1. The fuel flow rate is a manipulated input parameter to GTE and is regulated by a throttling lever. The pressure and temperature at each station and the shaft speed are considered output parameters of interest. Various operating modes have been considered by varying the fuel flow quantity to the engine over the entire operating range, and the engine data is logged. The analysis of the raw data used for training is given in Table 6.2.

Remark 1 *The assumption is made here that the engine is running in an ideal condition, and no malfunction occurs during the data collection process. This data dependency can also be considered one of the critical disadvantages of deep learning–based applications to engineering systems.*

6.4.2 Data Pre-Processing

Generally, the raw data collected from the experimental set-up consists of noisy sensor information and missing values. The pre-processing of data is an essential step that includes data cleaning and normalising.

1. Data filtering: The raw data is filtered through an exponential weighted moving average algorithm, which is also known as a first-order filter, to filter out the noise. This filter is represented by

TABLE 6.1
Sensor Signals of the Laboratory SR-30 Mini GTE Data Set

Index	Symbol	Description	Unit
1	P_1	Ambient pressure	kPa
2	T_1	Ambient temperature	K
3	$\dot{m}_f$	Fuel flow rate	L/Hr
4	P_2	Compressor exit pressure	kPa
5	T_2	Compressor exit temperature	K
6	P_3	Combustor exit pressure	kPa
7	T_3	Combustor exit temperature	K
8	P_4	Turbine exit pressure	kPa
9	T_4	Turbine exit temperature	K
10	N	Shaft speed	rpm

TABLE 6.2
Raw Data Properties in the Training Set

Parameters	P_1 (kPa)	T_1 (K)	$\dot{m}_f$ (L/Hr)	N (rpm)	P_2 (kPa)	T_2 (K)	P_3 (kPa)	T_3 (K)	P_4 (kPa)	T_4 (K)
Mean	110.31 3	308.12	18.25	62060.6	250.15	428.33	240.11	921.58	192.31	846.08
Std	3.01	0.56	1.87	7325.36	21.35	61.55	32.35	59.089	33.87	58.73
Min	105.08	306.73	12.95	48161.24	154.79	389.99	148.60	878.70	138.96	714.17
25%	107.72	307.76	16.72	56140.18	197.08	412.01	189.20	892.92	162.75	803.87
50%	108.98	308.07	17.97	59577.0	242.44	424.84	232.74	915.98	176.55	850.10
75%	113.38	308.5	20.06	69703	302.76	446.8	290.65	944.97	225.21	896.04
Max	115.72	309.83	20.45	73905.65	368.33	469.25	353.59	993.77	253.63	936.31

$$\hat{x}_i = \lambda x_i + 1 + \lambda x_{i-1}$$

where λ is a filter factor that ranges from 0 to 1; x_i and x_{i-1} are the consecutive samples from the raw data; and $\hat{x}$ is the filtered data.

2. Data normalisation: The features of the GTE parameters selected for DL applications are expressed in different units. Also, the order of magnitude of their values is different, e.g. shaft speed range is 45000–74000 rpm, whereas the fuel flow rate range is 1–25 L/Hr. Hence, a linear transformation is applied for normalising the featured data, which is given as

$$\tilde{x} = \frac{\hat{x} - \mu}{\sigma}$$

where $\hat{x}$ is the filtered data set; μ is the average; and σ is the standard deviation.

6.4.3 Configuration of Network

This step includes the network configuration and the selection of hyper-parameters such as the optimiser, learning rate, and minimum batch size.

1. First, the pre-processed data is partitioned into two sets: 70% of the data is featured at random for training and 30% is kept for testing. In addition, 15% of the data from the training set is held for validation to understand the network behaviour and the generalisation ability on the unseen data.
2. The LSTM network is configured by choosing 40, 30, 20 neurons in three different hidden layers, respectively, which are specified in Figure 6.5 to create a deep neural network.
3. The cost function is chosen to be the root mean squared error function as

$$\text{RMSE} = \sqrt{\frac{1}{n}\sum_{i=1}^{n}\left(\hat{\boldsymbol{y}}^{(i)} - \tilde{\boldsymbol{x}}^{(i)}\right)^2} \tag{6.10}$$

where i is the index of the sample; $\hat{\boldsymbol{y}}$ is the predicted response; $\tilde{\boldsymbol{x}}$ is the measured response of the parameters from Step 4.2; and n is the sample size in the data set.

4. Lastly, a learning algorithm is selected for the mentioned framework, Adam [21], an optimisation scheme for minimising the performance index in the LSTM network. The learning algorithm learns the features iteratively by minimising the cost function with every iteration by selecting a correct combination of weights and biases.
5. Hyper-parameter tuning is done based on the feedback received from the performance analysis on the validation data set as given in Table 6.3.

It is essential to point out here that an *epoch* is different from an *iteration*. One epoch refers to a single pass of the entire training data set in the forward and backward direction

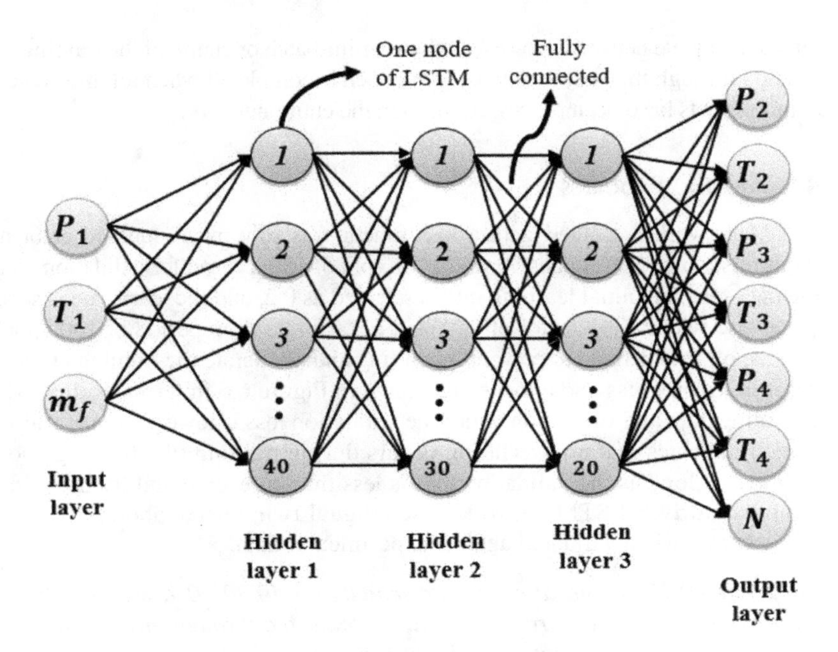

FIGURE 6.5 LSTM network–based digital twin of GTE.

TABLE 6.3

Set of Hyper-Parameters to be Tuned during Training

Field	Description	Value
Epoch	Controls the number of complete passes of the training data set through the network	500
Mini-batch size	Updates the weights in the next iteration by calculating the gradient of the loss function on a piece of the training data set	32
Validation frequency	The number of iterations between evaluations of validation metrics	20
Base learning rate	A small positive value that controls the step size in which the weights are updated during training	0.2
L_2 regularisation	Useful when the training data set consists of collinear/co-dependent features that improve the prediction performance and reduce overfitting	0.001

through the complete network. However, when a mini-batch or chunk of the training data set is passed through the neural network. An epoch is completed when all mini-batches of the training sets have been processed through the entire network.

6.4.4 Training Progress

The LSTM network is trained with a loss function as root mean squared error and Adam optimiser over 500 epochs. A validation split of 15% is used for validation on the training data set. The initial learning rate is selected as 0.2, and the learn rate is scheduled piecewise with a drop factor of 0.01 and a learn rate drop period of 100 epochs. A batch size of 32 is defined to prevent overfitting and accelerate the training process.

The training progress and epochs are shown in Figure 6.6 in terms of the RMSE of the training data and validation data. The validation loss is lower than the training loss, which indicates that no overfitting occurs during training (the training should be continued as long as the validation loss is less than or even equal to the training loss). The data-driven LSTM network–based digital twin of the laboratory GTE is ≈92.5% accurate when validated against experimental data.

Remark 2 *The DNN configuration is represented as 3:40:30:20:7, which means that the network consists of three inputs, fully connected through three hidden layers with 40, 30, and 20 neurons in each layer, respectively, seven outputs predicted for the engine output responses.*

6.4.5 Testing and Deployment

The final step in the LSTM network–based digital twin is testing the trained network on a new data set. Once the performance is found satisfactory, the weights and biases can be

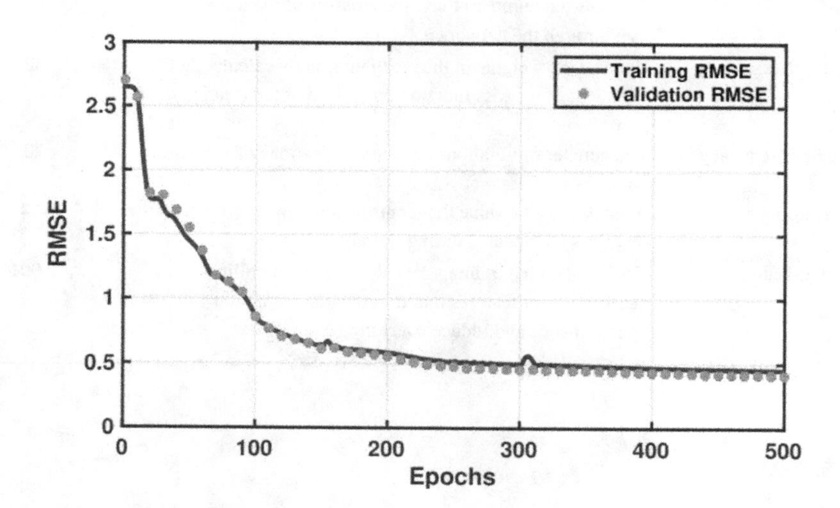

FIGURE 6.6 Progress of training RMSE through the epochs.

frozen, and the trained network can be deployed on edge devices or a hardware unit to predict the dynamic behaviour of engine parameters for a given input pattern. It is to be pointed out that the experimental data is logged in the LabVIEW interface and is read by Matlab software via open platform communication (OPC). The OPC toolbox provides access to live data directly to Matlab from LabVIEW. Here, Matlab serves as a client and collects live data that is fed as input to the LSTM network–based digital twin. The output response predicted by networks is compared against the measured data from the system. This assembly constitutes the National Instrument Data acquisition system, LabVIEW interface, and Matlab toolbox to perform hardware-in-loop model validation developed by the proposed deep learning framework.

6.5 RESULTS AND DISCUSSION

This section includes the simulation and validation response obtained from the LSTM network–based digital twin. A set of experiments collects a huge experimental data set to perform validation of the developed digital twin. A LabVIEW interface is used for logging GTE sensor measurements at each station. The obtained data is then used to validate the proposed data-driven LSTM network–based digital twin. The performance of the LSTM-based digital twin is compared with that of the NARX neural network; the architecture is described in Appendix A.1. For the NARX-based model, seven different data-driven networks are configured, trained, and tested. Further, the performance of the LSTM-based digital twin is compared with that of a first principle–based mathematical model (FPM) described in Appendix A.2.

6.5.1 Model Validation on Testing Data Set

First, the trained LSTM network is validated on the testing data set. The experimental input from the testing data set is fed as input to the trained LSTM network, and the predicted response is obtained in the form of non-dimensional output. It is always suggested

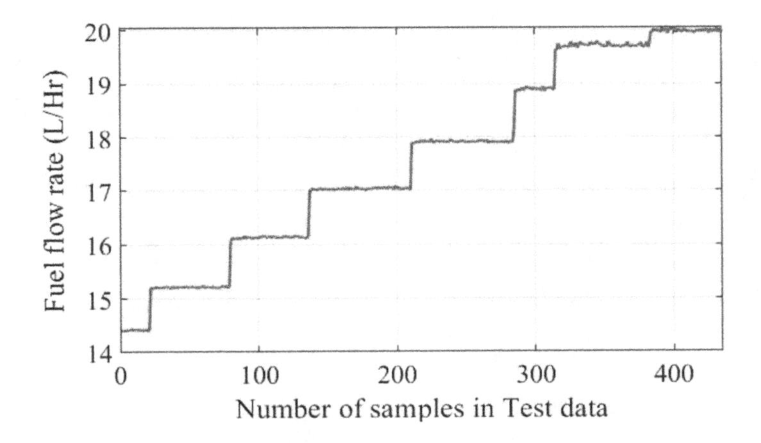

FIGURE 6.7 Test set input to the LSTM network.

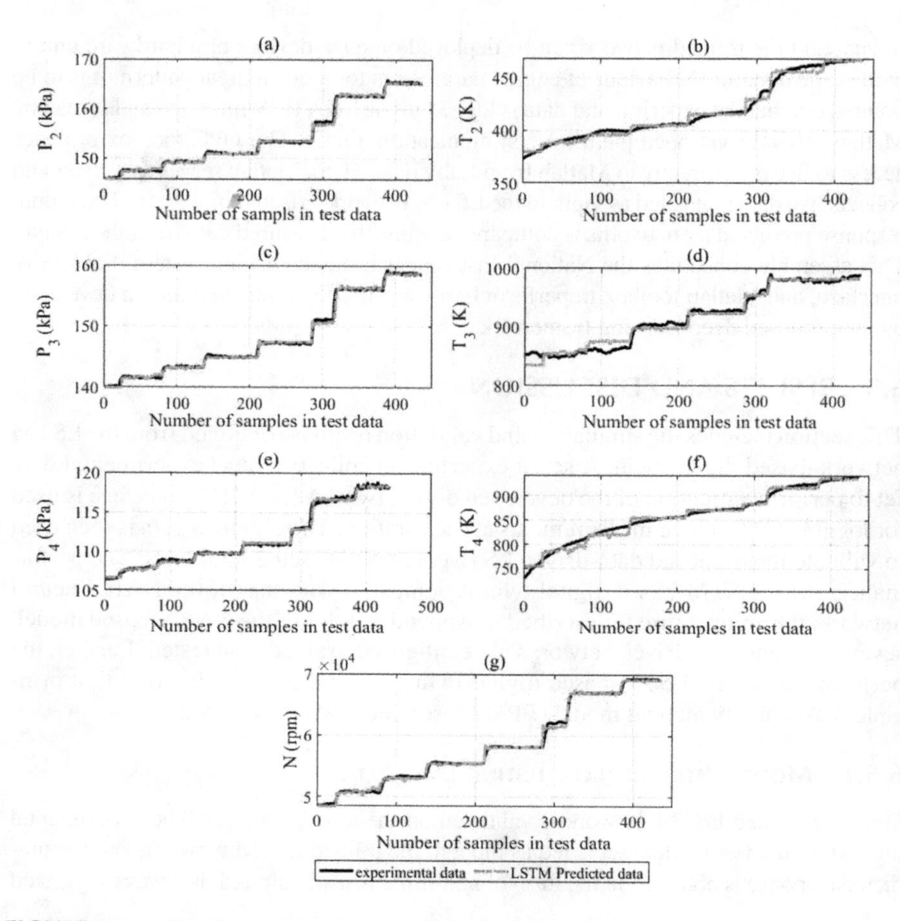

FIGURE 6.8 LSTM network validation on the testing data set.

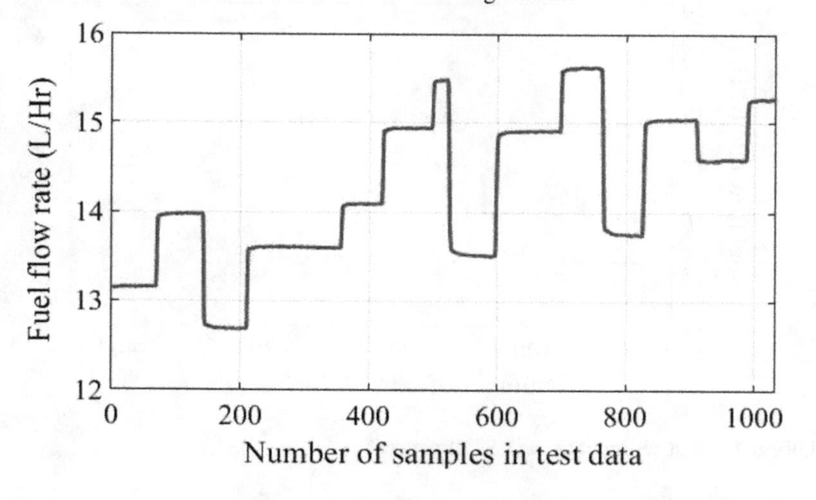

FIGURE 6.9 Experimental fuel flow rate as input to LSTM and NARX networks.

to compare the predicted output of the dynamical system in its original units. Hence, first, the predicted output is denormalised from the training set information, and the final outcome is shown against the experimentally measured output. For the given input signal as shown in Figure 6.7, the predicted responses and measured outputs are shown in Figure 6.8. The predicted responses of the dynamic parameters are in close agreement (≈97.25%) with the measured output from the experimental set-up.

6.5.2 LSTM Network Validation against NARX Network

The LSTM network–based digital twin is then used to validate the real-time data set. Later, the experimental set-up input is fed to the trained LSTM network–based digital twin. Finally, the output response predicted by the LSTM network–based digital twin is compared against a conventional NARX network–based model. Since a single

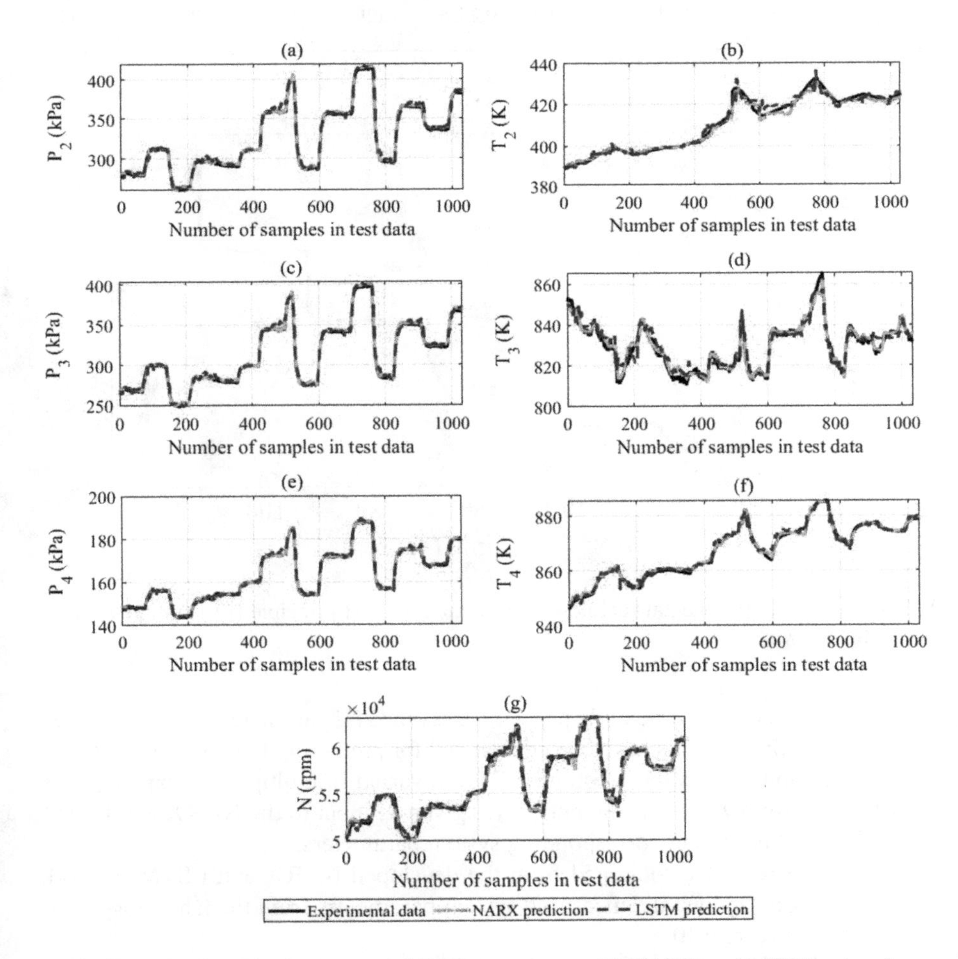

FIGURE 6.10 Predicted response of GTE parameters with NARX and LSTM networks.

TABLE 6.4
Performance Comparison of LSTM and NARX Networks in Terms of RMSE

Parameter →	N	P_2	T_2	P_3	T_3	P_4	T_4
NARX network prediction performance							
Network configuration	3:5:8:1	3:5:4:1	3:8:8:1	3:5:3:1	3:8:11:1	3:5:6:1	3:5:4:1
Training time (sec)	1220	413	840	424	1512	537	1040
Training error	0.1256	0.0930	0.1001	01077	0.2286	0.0891	0.0882
Testing error	0.1147	0.0918	0.0913	0.1018	0.2126	0.0885	0.0837
LSTM network prediction performance							
Network configuration	3:40:40:30:7						
Training time (sec)	3072						
Training error	0.2677	0.1404	0.2545	0.1402	0.4266	0.1509	0.2083
Testing error	0.2405	0.1299	0.2536	0.1295	0.4200	0.1354	0.2008

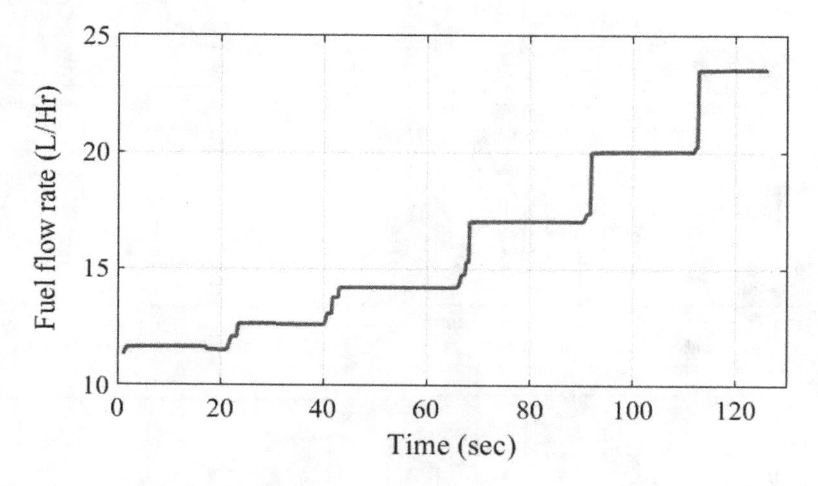

FIGURE 6.11 Experimental fuel flow rate as input for first principle (FP) and the LSTM-based digital twin.

NARX network cannot efficiently predict the seven GTE parameter responses, seven different NARX networks have been developed for predicting the response.

The fuel flow rate to the laboratory GTE is varied in multiple random steps, as shown in Figure 6.9. This new input data is given as input to the NARX and LSTM network model to predict corresponding system parameters.

The output responses obtained from the developed NARX and LSTM network model compared against the measured system parameters from the laboratory GTE are shown in Figure 6.10.

The predicted response obtained indicates that both networks offer good accuracy in predicting the engine parameters' dynamic behaviour. The performance measures for

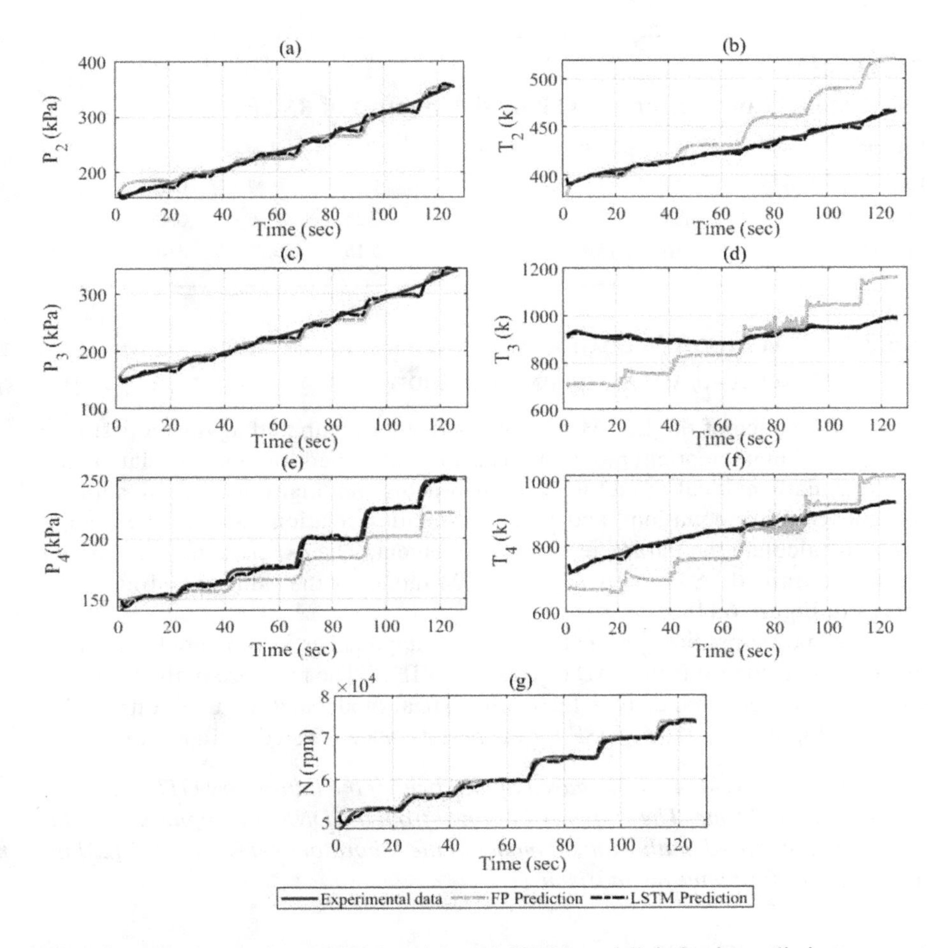

FIGURE 6.12 Comparison of FPM model and LSTM-based digital twin prediction response of GTE parameters.

both frameworks are illustrated in Table 6.4 in terms of the RMSE calculated between the predicted output and the expected output. The prediction accuracy of the NARX network–based data-driven model is ≈96% by averaging the error over the considered seven networks. The NARX network is capable of extracting dynamic behaviour when dealing with single-input and single-output. However, we needed seven different networks to predict the dynamic response. As has been tested, a single network is not efficient when mapping with seven outputs simultaneously. The LSTM network is quite efficient in the time-series prediction of engine behaviour. The data-driven LSTM network–based digital twin is found to be ≈92.5% accurate when validated against the experimental data set. However, it can be pointed out that the three different gates that characterise a cell of LSTM have their own set of operations, thereby making the training speed slower than that of the NARX network.

TABLE 6.5
LSTM Model Comparison with FP model in terms of RMSE

Parameter	N	P_2	T_2	P_3	T_3	P_4	T_4
LSTM model	0.74	1.864	0.364	1.87	0.47	0.52	0.48
FPM model	0.83	2.16	0.45	2.23	1.16	0.45	0.76
Improvement (%)	12.16	13.70	19.11	16.15	59.79	13.46	36.46

6.5.3 LSTM Model Validation against a First Principle–Based Mathematical Model

The performance of the LSTM network is further compared against a first principle–based mathematical model of GTE [5], developed for control-related studies. The mathematical model is developed using modular concepts by applying the conservation equations and thermodynamics relations among the components to calculate the parameters of each element. The experimental fuel flow is fed to the trained LSTM network and FPM model at the same sampling rate as shown in Figure 6.11.

The response predicted by both modelling approaches is compared against the measurements logged from DAQ through LabVIEW. The response of the laboratory engine parameters predicted by the mathematical model and the LSTM network is shown in Figure 6.12. The RMSE for the performance is listed in Table 6.5.

Remark 3 *The LSTM network–based digital twin of the laboratory GTE is then validated in real-time. The response obtained from the DNN is compared with the first principle–based mathematical model of the laboratory GTE. See Ref. [22] for a detailed performance comparison*

6.6 CONCLUSIONS

We proposed an LSTM network–based digital twin for predicting the dynamic responses of engine parameters. The predicted responses indicate that the LSTM network–based digital twin delivers high accuracy in predicting the dynamic response of GTE parameters with proper training and the selection of hyper-parameters. The LSTM network predicts seven outputs with an accuracy of $\approx$92.5% through a single network. Although the NARX model provides a more accurate prediction, it requires seven separate neural network models for each system parameter, which is a cumbersome process. Experimental validation results also confirm that an LSTM network–based digital twin can be considered a suitable alternative to first principle–based models. The proposed DNN model is regarded as a digital twin of the laboratory set-up, utilised to monitor the actual set-up's performance and provide virtual sensor measurement in case of any sensor failure. Further research work is planned in the area of engine health monitoring, fault detection, and fault-tolerant control using intelligent techniques.

APPENDIX 6A

6A.1 NARX Neural Network Architecture

The autoregressive exogenous (ARX) is a standard system identification method. NARX is the non-linear generalisation of the ARX model. It is essentially a time-series network consisting of multiple layers of neurons and a non-linear function just before the output layer. They join together in a feedforward manner and feedback connections with a tapped delay line to enclose several layers. The output signal ($\hat{y}(k+1)$) is predicted by the previous value of output ($\hat{y}(k)$) and exogenous (independent) inputs of a time-series data set [1, 23].

In this network, the predicted output, represented as $\hat{y}(k)$, is fed back to the input of the feedforward neural network as part of the standard NARX structure to yield

$$\hat{y}(k+1) = f\left(\hat{y}(k), \hat{y}(k-1), \ldots, \hat{y}(k-d_y), u(k-1), u(k-2), \ldots, u(k-d_u)\right) \tag{6A.1}$$

where $\hat{y}(k)$ is the predicted response from the neural network, which is an estimate of the actual system output $y(k)$; $u(k)$ is the system external input; d_y and d_u represent the input and output delays, respectively; and f denotes the non-linear mapping function. During training, a conventional multilayer perceptron (MLP) network approximates the mapping function $f(v)$, which is initially unknown.

Similarly, the multi-step ahead of output prediction for *n-step* is given as

$$\hat{y}(k+n) = f\left(\hat{y}(k-d_y), \hat{y}(k-d_y-1), \ldots, \hat{y}(k), u(k-d_u), u(k-d_u-1), \ldots, u(k)\right) \tag{6A.2}$$

The NARX network has found application in prediction, noise filtering, and dynamic modelling. The detailed architecture and variations of the NARX network are given in Ref. [1] (Figure 6.13).

6A.2 First Principle–Based Mathematical Model of Laboratory GTE

A control-oriented, first principle model is developed on the modular concept, simulated in Matlab to predict the dynamic response of a GTE. The model dynamics are derived using the conservation equations and thermodynamical relations among the engine components. The FPM model consists of five subsystems, and each subsystem addresses an individual part of the engine, with its own set of thermodynamic relations and performance map data set. A hybrid modelling approach, the "state variable method", is proposed to develop a mathematical model that captures the steady-state and transient dynamics of a laboratory GTE. The system dynamics is represented by a set of three first-order differential equations that are defined by

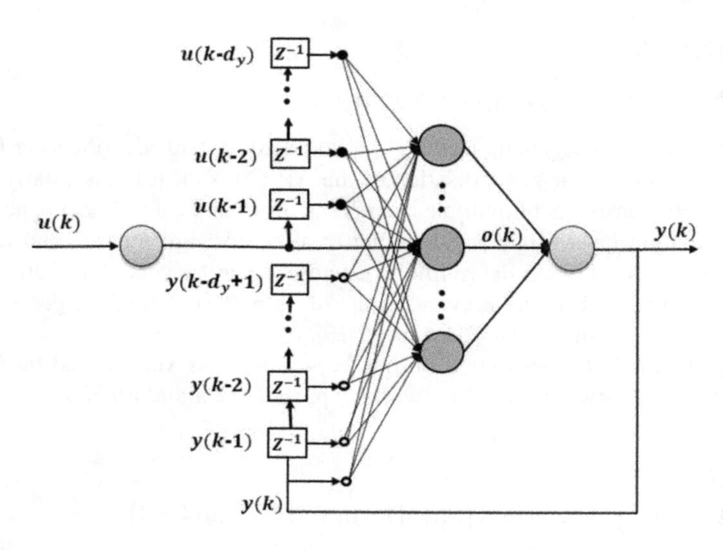

FIGURE 6.13 NARX network with tap delay layer.

associating with each state variable, i.e. $[P_2P_4N]$. The mathematical model of a GTE can be expressed in non-linear dynamical form as

$$\dot{x} = f\left(\boldsymbol{x}, u, \boldsymbol{w}_p\right) \tag{6A.3}$$

where $x = [P_2P_4N]$ is the system state vector; P_2 is the compressor exit pressure; P_4 is the turbine exit pressure; N is the shaft speed; and $u = \dot{m}_f$ is the control input. The vector $\boldsymbol{w}_p$ is a set of parameters obtained from component maps (an engine performance map) that are developed using experimentation tests [5,6]. A detailed description of the mathematical modelling of each component, including steady-state and dynamic analysis, is presented in Ref. [5]. This mathematical model of the laboratory GTE is later used to implement and verify the various control design approaches.

REFERENCES

1. Asgari, Hamid, and XiaoQi Chen. 2015. *Gas Turbines Modeling, Simulation, and Control: Using Artificial Neural Networks.* New York: CRC Press.
2. Patel, Nikunj. 2019. How aerospace industry can use digital twins to improve fleet management and sustainment. www.einfochips.com/blog/how-aerospace-industry-can-use-digital-twins-to-improve-fleet-management-and-sustainment/.
3. Schobeiri, M. T., M. Attia, and C. Lippke. 1994. Getran: A generic, modularly structured computer code for simulation of dynamic behavior of aero and power generation gas turbine engines. *Journal of Engineering for Gas Turbines and Power*, vol. 116, no. 3, pp. 483–494.
4. Crosa, G, Pittaluga, F., Trucco Martinengo, A., Beltrami, F., Torelli, A., and Traverso, F. 1998. Heavy-duty gas turbine plant aero-thermodynamic simulation using Simulink. *Proceedings of the ASME 1996 Turbo Asia Conference*, Jakarta, Indonesia, pp. 550–556.

5. Singh, Richa, Arnab Maity, and P. S. V. Nataraj. 2018. Modeling, simulation and validation of mini SR-30 gas turbine engine. *IFAC Proceeding in Advances in Control and Optimization of Dynamical Systems*, Hyderabad, India, vol. 51, pp. 554–559.
6. Rowen, William I. 1983. Simplified mathematical representations of heavy-duty gas turbines. *Journal of Engineering for Power*, vol. 105, no. 4, pp. 865–869.
7. Kim, J. H., T. W. Song, T. S. Kim, and S. T. Ro. 2001. Model development and simulation of transient behavior of heavy duty gas turbines. *Journal of Engineering for Gas Turbines and Power*, vol. 123, no. 3, pp. 589–594.
8. Surendran, Swathi, Ritesh Chandrawanshi, Sanjeet Kulkarni, Sharad Bhartiya, Paluri SV Nataraj, and Suresh Sampath. 2016. Model predictive control of a laboratory gas turbine. *Indian Control Conference (ICC)*, Kanpur, India, pp. 79–84, IEEE, 2016.
9. Kalogirou, Soteris A. 2001. Artificial neural networks in renewable energy systems applications: A review. *Renewable and Sustainable Energy Reviews*, vol. 5, no. 4, pp. 373–401.
10. Basso, M., F. Bencivenni, Laura Giarre, S. Groppi, and G. Zappa. 2002. Experience with NARX model identification of an industrial power plant gas turbine. In *Proceedings of the 41st IEEE Conference on Decision and Control*, vol. 4, pp. 3710–3711.
11. Bettocchi, Roberto, Michele Pinelli, Pier Ruggero Spina, Mauro Venturini, and Michele Burgio. 2004. Set up of a robust neural network for gas turbine simulation. In *Turbo Expo: Power for Land, Sea, and Air*, Vienna, Austria, vol. 41693, pp. 543–551.
12. Kaiadi, Mehrzad. 2006. Artificial Neural Networks Modeling for Monitoring and Performance Analysis of a Heat and Power Plant. Master's thesis, Division of Thermal Power Engineering, Department of Energy Sciences, Lund University, Sweden.
13. Punjani, Ali, and Pieter Abbeel. 2015. Deep learning helicopter dynamics models. *International Conference on Robotics and Automation (ICRA)*, Seattle, USA, pp. 3223–3230. IEEE, 2015.
14. Mohajerin, Nima, Melissa Mozifian, and Steven Waslander. 2018. Deep learning a quadrotor dynamic model for multi-step prediction. *International Conference on Robotics and Automation (ICRA)*, Brisbane, QLD, Australia, pp. 2454–2459. IEEE, 2018.
15. Palagi, Laura, Apostolos Pesyridis, Enrico Sciubba, and Lorenzo Tocci. 2019. Machine learning for the prediction of the dynamic behavior of a small scale ORC system. *Energy*, vol. 166, pp. 72–82.
16. Xu, Yan, Yanming Sun, Xiaolong Liu, and Yonghua Zheng. 2019. A digital-twin-assisted fault diagnosis using deep transfer learning. *IEEE Access*, vol. 7, pp. 19990–19999.
17. LeCun, Yann, Yoshua Bengio, and Geoffrey Hinton. 2015. Deep learning. *Nature*, vol. 521, no. 7553, pp. 436–444.
18. Yu, Xinghuo, M. Onder Efe, and Okyay Kaynak. 2002. A general backpropagation algorithm for feedforward neural networks learning. *IEEE Transactions on Neural Networks*, vol. 13, no. 1, pp. 251–254.
19. Graves, Alex., ed. 2012. Long short-term memory. In *Supervised Sequence Labelling with Recurrent Neural Networks*. Berlin, Heidelberg: Springer, pp. 37–45.
20. *Gas Turbine Power System Operator Manual*. 2005. Turbine Technologies Ltd. Minilab, 410 Phillips Court, USA.
21. Kingma, Diederik P., and Jimmy Ba. 2015. Adam: A method for stochastic optimization. *3rd International Conference on Learning Representations, ICLR 2015*, San Diego, USA, May 7–9, 2015, Conference Track Proceedings.
22. Singh, Richa. 2018. Dynamic modeling of mini SR-30 gas turbine engine using deep learning. *MATLAB Expo: April 2018*.
23. Ng, Boon Chiang, Intan Zaurah Mat Darus, Hishamuddin Jamaluddin, and Haslinda Mohamed Kamar. 2014. Dynamic modeling of an automotive variable speed air conditioning system using nonlinear autoregressive exogenous neural networks. *Applied Thermal Engineering*, vol. 73, no. 1, pp. 1255–1269.

7 A Case Study of Additive Manufacturing in Prosthesis Development in Industry 4.0

M. C. Murugesh, R. Suresh, Ajith G. Joshi, and Priya Jadhav

CONTENTS

7.1 INTRODUCTION

The potential of additive manufacturing (AM) lies in the flexibility of component design and customisation. In AM, computer-aided design (CAD) models are developed based on individual patient needs (Altıparmak et al. 2021). Accordingly, the design and development of patient-specific models aid clinicians in planning and practicing medical procedures such as surgery, implants, diagnosis, and other anatomical models. Research in this area is progressing using a systematic approach towards widening the scope of AM for medical applications (Chang, Chang, and Lu 2021). One of the most important aspects of Industry 4.0 is 3D printing. Insights into work addressing the optimisation of process parameters, component cost reduction, better quality, and design freedom are summarised with an analysis.

DOI: 10.1201/9781003200857-7

7.2 PROSTHETIC LEGS

People with leg amputations can have normal function with the help of prostheses, which can have the appearance of real legs. A prosthesis is a replacement for an organic limb that may have been lost due to injury or deformation, and the mechanism of a prosthesis is designed to substitute for the function of the amputated limb. Part of a prosthetic limb is the residual limb, which refers to the part of the body that remains following an amputation. The residual limb, which is called the stump, is joined to the prosthesis by a socket (Radosh et al. 2017). The remaining part of the prosthesis is the adaptor which is aligned to the socket. The main function of the adaptor is to correctly assemble the foot below the leg. By studying the patient's walk and posture, especially walking on ground, the total length of the limb, and the balance of the limb during walking, adjustments can be made to the set-up which will allow the patient to walk easily and in a normal way. The underside of the adaptor is connected to the pylon. The pylon is a metallic part with a hollow cross section usually made of stainless steel or titanium that is connected to the socket of the prosthetic foot (Jin et al. 2015).

Adjustments can be made based on the overall size of the amputation and the measured height of the patient's leg. Due to technological advancements, progress in the design aspect of the pylon has changed from an easy static type to a dynamic type that provides rotational movement and also absorbs as well as stores energy. The fundamental role of the pylon is to act as a substitute for the length of the amputated fibula and tibia. The last part of the prosthesis is the base called the foot. The purpose of the foot is to provide load bearing and to absorb shock, replace lost muscle function, duplicate the lost muscle activity, and duplicate the anatomical joint plus reinstitute the cosmetic look (Pirouzi et al. 2014).

7.3 TREATMENT TIMELINE

Following an amputation, the patient requires substantial body therapy, and mental as well as emotional support. During this time, the patient has to learn how to use the prosthesis and attempt to return to their normal way of life. The stages are indicated in Figure 7.1. However, the timeline for recovery can vary due to the cost of a prosthesis and factors such as infection and reamputation (Rajťúková et al. 2014).

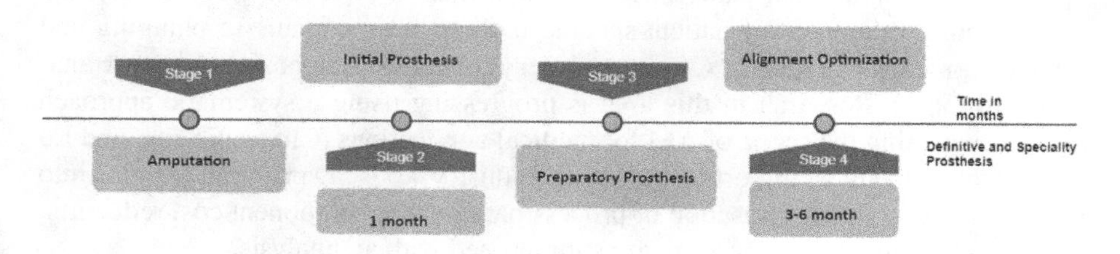

FIGURE 7.1 Timeline of prosthesis treatment.

The traditional prosthesis fabrication method uses metallic parts that are heavier and can lead to discomfort. The measurement process is time-consuming, and adding the fabrication time required for machining the complex geometries of metallic parts increases the time taken. Using additive manufacturing reduces the weight, and complex shapes can be manufactured with relative ease within a short time. A limb prosthesis manufactured using the traditional method incurs a high cost at the fabrication stage. Using additive manufacturing reduces the cost. 3D printing technology has revolutionised the world of manufacturing technology and it is widely used in the medical field as it facilitates the fabrication of complex geometries with ease and within a shorter time frame (Soud et al. 2019).

7.4 GENERAL PRESCRIPTION GUIDELINES

Another factor to be considered when a new implant is prescribed is the weight bearing of lower limb transtibial prostheses. If the patient has sensitive areas, scars, or neuromas, special considerations must be given to the socket and pylon design. In the pylon component, significant impact-absorbing materials should be employed, and changes may be required to transfer the weight across a wide area. A person who only uses the prosthesis indoors has different concerns and requirements than someone who expects to be active in their work and sports. Weight bearing, suspension, and structural strength/integrity are all affected by the extent of prosthesis use. The weight of the amputee's body, as well as their functional goals, must be matched with their usage level. The vast and ever-increasing number of prosthetic componentry options demand extensive consultation with a prosthetist. Different patients have distinct characteristics that must be assessed and factored into the prosthetic design. A person who lives near the sea, for example, may require a prosthesis that gives the best protection against salt corrosion and water damage; a carpenter, on the other hand, may want a prosthesis that provides more comfort in the kneeling position than the usual wearer. The cultural background of the amputee is also significant. During the manufacturing process, such personal aspects must be taken into account.

7.5 LITERATURE REVIEW

The article by Fariborz and Camila focuses on the key points that make additive manufacturing useful compared to the utility of traditional prosthesis fabrication techniques. Polylactic acid polymer materials to be used for developing prosthetic devices using additive manufacturing (Tavangarian et al. 2019). The article indicates that additive manufacturing can be less expensive than traditional methods. Also, the strength criteria of polymers and alloys are substantial and need to be considered when a lower limb prosthesis is required. To fabricate a prosthetic device using additive manufacturing, computer simulation software is used extensively as most design software have a simulation option. Additionally, using a computer simulation to nail down a design reduces the quantity of material wasted during physical testing. Furthermore, prosthetic devices can be modified to improve their performance through simulated testing.

7.6 RAW MATERIALS USED

Because a prosthetic device should be feathery and light, it is mostly made of plastic. Polypropylene is commonly used for the socket. Formerly, a pylon was generally made of steel, but nowadays steel has been replaced with lightweight metals such as titanium and aluminium (Jadhav et al. 2021). Alloys of the aforementioned materials are most frequently used. Most recently, carbon fibre and its reinforced materials are being used to form a lightweight pylon. However, a pylon made of carbon fibre is very expensive. Rubber and wood from trees such as maple, hickory basswood, willow, poplar, and linden have traditionally been used for limb sections such as the foot. Recently, they have been made from urethane foam with a wood inner keel. Plastics such as polyethylene, polypropylene, acrylics, nylon, silicones, polyurethane, and rigid polyurethane foam are also often employed. Soft yet durable materials are used to make prosthetic socks. Fibre-reinforced glass and carbon are used for their lightweight yet high strength characteristics. Their only drawback is that they are expensive and fairly unaffordable for the majority of amputees. A soft polyurethane foam cover is meticulously designed to fit the form of the patient's sound limb and conceal the majority of the pylon. The foam cover is then painted with a sock or artificial skin to make the patient's skin look better (Wu et al. 1979).

7.7 PROSTH ESIS BY ADDITIVE MANUFACTURING OVER CONVENTIONAL METHODS

The convenience and adaptability of a prosthesis are essential to consider. According to a survey, nearly 14.3% of amputees discard their prostheses because they are confining or hard to wear, and these characteristics exceed their advantages (Kumar et al. 2020). Amputees suffer aches and pain from high contact pressure points generated by the pylon. The residual limb's volume change is another challenge that is the consequence of high pressure points generated by the pylon. The pylon, which is usually composed of lightweight but high strength materials such as aluminium and titanium, generates high amounts of reaction forces back to the residual limb, which generate pressure points directed straight up to the stitches made on the residual limb during surgery. This not only creates high pressure points but also causes the amputee intense pain. Also, the high pressure points created by the pylon trigger a huge volume change in the tissues and tendons situated around the bony architecture of the knee. The change in pressure and pressure points is associated with the volume change.

In addition to aches and pain, these high pressure points also deteriorate the quality and life of the socket and will compel the amputee to use several sockets over time. The problems arising from the use of a hard pylon part can be eradicated using 3D printing, where materials such as polylactic acid (PLA) and Kevlar are sufficiently soft such that they reduce pressure points, and are lightweight yet strong enough to withstand the weight of the amputee. The additive manufacturing of prosthetic components would help address some of the challenges that amputees encounter such as cost, the recurring need for newly sized prosthetics, and the preference to have working and practical prosthetics. Computer-aided design programs, 3D printing, and various image software can provide an inexpensive and attainable solution. Magnetic resonance imaging (MRI) and computer tomography (CT) can be used to construct a prototype of the amputee's stump. Laser scanners can record the limb contours and render them in a CAD program (Song et al. 2019). These technologies can be implemented to boost 3D printing and shorten the process of creating impressions and negative and positive moulds for socket preparation. Editing image programs can be used to assess dimensions and make measurements. Further, various software can be incorporated to scale the prosthetic to the desired size. Inexpensive prosthetic devices would greatly enhance amputees' quality of life, regardless of their socio-economic circumstances. 3D printing helps reduce the rescaling costs compared to previously used methods.

Quick implementation and fabrication are the main advantages of 3D printing as it takes less time and it is also easier to reprint, update, and modify a prosthesis (Paital and Dahotre 2009); the more information available on the patient, the faster the process. The conventional process, on the other hand, can take from a few weeks to several months to create a standard prosthesis and this depends on the medical structure and its availability. In regions with scarce resources, patients go through an extended journey and significant expense to reach a prosthetist, as many are based in big cities. Following the first medical examination, many patients try for other solutions with the consequence that the resultant prosthesis might escalate the possibility of the body rejecting the prosthesis treatment.

The current prosthetic device fabrication process is work-intensive. During fabrication, plaster moulds and extra fabrication materials are discarded, which might be termed a waste of materials. The entire technique must be performed in the event that additional sockets are required due to inherent residual limb changes. The prosthetist's preliminary inspection and experience define the position and thickness of build-ups to adjust the positive plaster or foam mould for establishing efficient plumb points, which vary from patient to patient and prosthetist to prosthetist. If the modifications are incorrect, the whole prosthetic mould has to be recreated, and the previous mould is discarded as waste. A pylon must be fabricated with accurate dimensions as there is no possibility of adding material if the dimensions are incorrect. In the latter case, the pylon has to be recreated. In the case of 3D printing, the process itself is additive while the conventional process is subtractive. Changes and modifications to a pylon can be easily done, at any point in time.

7.8 FUSED DEPOSITION MODELLING

Fused deposition modelling (FDM) is an additive manufacturing process that involves the layer-by-layer depositing of melted material in a predefined route. The materials are thermoplastic polymers that come in the form of filaments (Yang et al. 2012). FDM is a 3D printing technology that is extensively used (Figure 7.2).

Spools are used as the material for the FDM process. The printer is initially loaded with a spool of thermoplastic filament. The filament is fed to the extrusion head and melts in the nozzle after the nozzle has achieved the desired temperature.

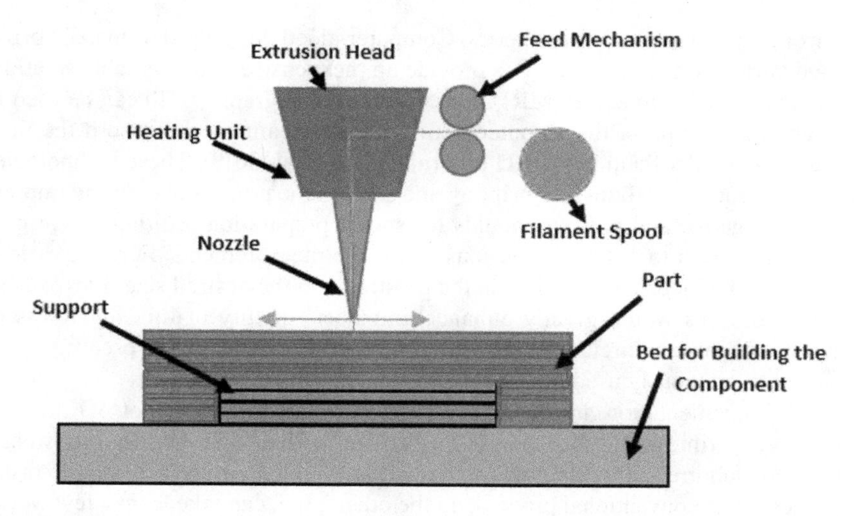

FIGURE 7.2 Additive manufacturing (FDM) process.

The extrusion head is then connected to a three-axis system, allowing it to move in three directions: X, Y, and Z. Extruded thin strands of melted material are inserted layer by layer in preset locations, where they cool and solidify. To speed up the cooling of the material, cooling fans connected to the extrusion head are occasionally utilized. Some areas will take multiple passes to fill. After finishing the printing layer, nozzle or extrusion head moves and deposits a new layer. This process is carried out until the part is completed. The thickness of FDM layers is typically between 50 and 400 microns. Smoother components and more accurate curved geometries arise from a lower layer height, while parts manufactured faster and at a lower cost result from a higher layer height. The most common layer height is 200 microns. Several process variables, including the temperature of both the nozzle and the build platform, the build speed, the layer height, and the cooling air flow, can be changed in most FDM systems. A home 3D printer's build size is typically 200 × 200 × 200 mm, while industrial machines will have a build size of 1000 × 1000 × 1000 mm.

7.9 CASE STUDY

Data collection from M S Ramaiah Limb Centre Gnanagangothri Campus, Bangalore, India studied the current manufacturing methods and processes of a prosthetic limb, the materials used during the processes, and the challenges faced in manufacturing a prosthetic limb. Finding the pressure points in each patient is challenging. The material used as a negative mould was plaster of Paris (POP). After filling the negative mould with POP, the actual prosthetic limb structure was obtained. The traditional process of fabricating a prosthetic limb is shown in Figure 7.3.

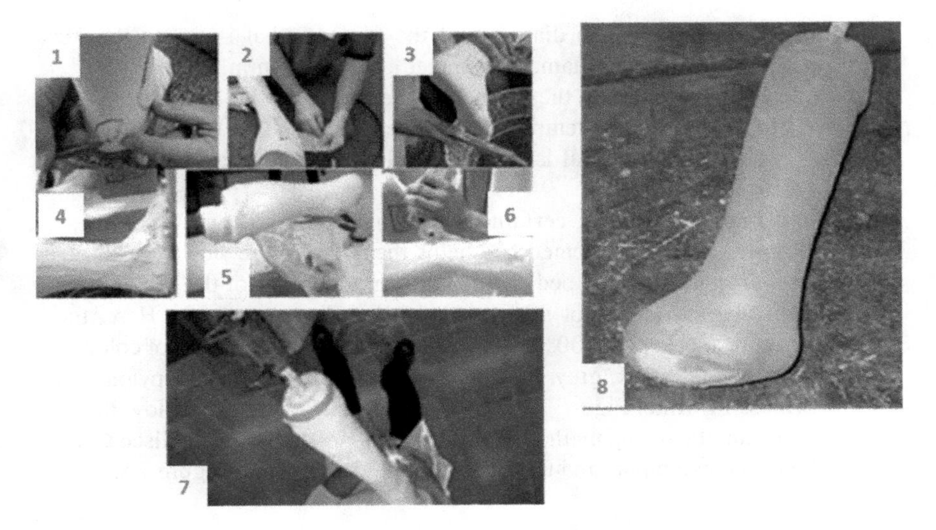

FIGURE 7.3 Traditional process of carrying out the fabrication of a prosthetic limb.

7.9.1 Development of Specification for Prosthetic Limb

Prosthetic limb is obtained from Germany and the data received from the limb centre, the following specifications were developed.

1. **Infill density:** The amount of plastic used on the print's interior. More plastic is used on the interior of the print with a higher or bigger infill density, resulting in a well-built design. In models with an aesthetic function, an infill density of about 20% is used; bigger or higher densities can be used for end-use components.
2. **Fabrication of pylon:** A pylon was fabricated using the fused deposition modelling technique with polylactic acid material. Since the pylon part is the end user part, the infill density was given as 100%.
3. **Layer thickness:** The thickness in 3D printing is the layer height of each succeeding insertion of material in the additive manufacturing or 3D printing process where layers are stacked. It is one of the necessary technical characteristics of any 3D printer; the layer height is actually the vertical resolution of the z-axis. Layer thickness is 150 microns/0.15 mm. The 3D printer extrusion width of each line in the print is directly affected by the nozzle diameter.
4. **Nozzle size:** The diameter of the nozzle that extrudes the material layer by layer is connected to the nozzle size. The nozzle diameter has a direct effect on the extrusion width of each line in a 3D print. The most popular standard nozzle size is 0.4 mm (or 0.35 mm) because it is an excellent all-round nozzle size.

5. **Filament diameter:** The diameter of the filament that is fed to the 3D printer heater core. The filament diameter used is 1.75 mm.
6. **Nozzle temperature:** As the filament is heated from the heater core and travels to the nozzle, the temperature of the nozzle should be around 180–230°C so that the material is finely extruded. Here, the temperature was kept at around 230°C.
7. **Heated bed:** On cooling, certain fibres contract and distort. When these plastics are extruded, a heated bed helps them cool more slowly, resulting in less warping. A heated bed ensures proper adherence of the initial layer and that the item does not fall out of the bed during printing. Here, the temperature was kept at 600°C. PLA is available in a wide range of colours, including white, black, grey, blue, and red. For this project, the pylon was 3D printed in white. Figure 7.4 shows a flow chart of the work flow in the additive manufacturing method of a lower limb prosthesis. A finalised CAD model of the pylon part and the truss elements is shown in Figure 7.5.

7.9.2 Finite Element Analysis of Pylon

Assumptions made for the analysis:

- Geometry model considered for FEA is a quarter symmetry model.
- One end is fixed and the equivalent load is applied at the other end.
- The complete structure is considered as a single structure (no welds or riveted joints).

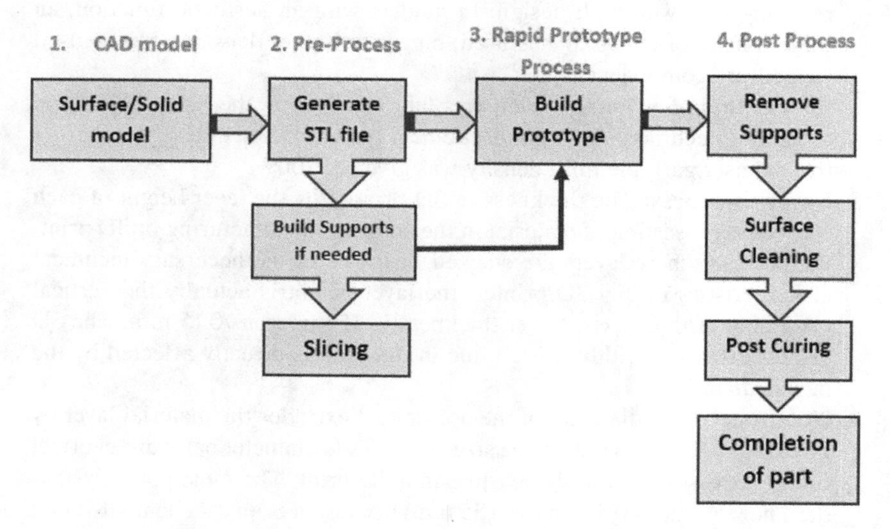

FIGURE 7.4 Flow chart of the work flow in the additive manufacturing method of a lower limb prosthesis.

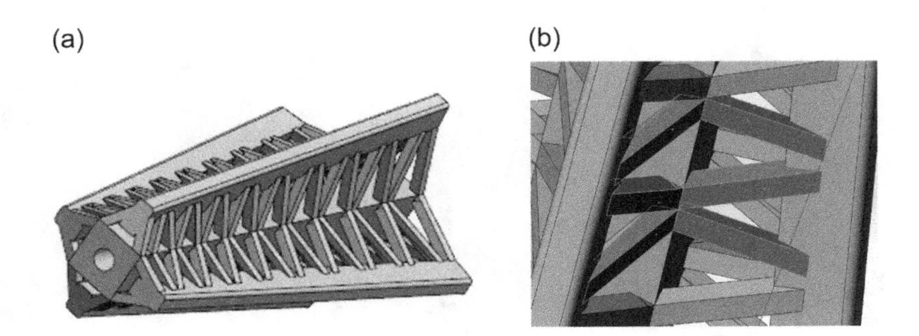

FIGURE 7.5 Finalised CAD model of (a) the pylon part and (b) the truss elements. (Dimension of the pylon: Length from top to bottom: 270 mm, internal diameter of the frame 12 mm, thickness of the frame 20 mm.)

A geometric model of a pylon considered for FEA is shown in Figure 7.6 and loads are applied on the square cross section of the pylon as in Figure 7.7.

Two rows of nodes on top plate constrained in the Z-direction, Cut section face nodes are constrained in Symmetry X and Z direction, Bottom face nodes are constrained in Y-direction

Figure 7.8 shows the Von Mises stress distribution (values in MPa) in the pylon. The maximum yield stress is 5.15 MPa. The maximum displacement (0.182 mm) occurs at the top portion of the pylon (where the socket will be fitted) and the minimum displacement (0.02 mm) occurs at the bottom portion (where the foot will be fitted). The final assembly of a 3D printed below knee prosthesis with various attachments is shown in Figure 7.9.

The following are the key categories where user-friendly and long-lasting implants are required. In some areas, there is a greater availability of skilled professionals who can give patients high quality, custom-made prosthetic devices.

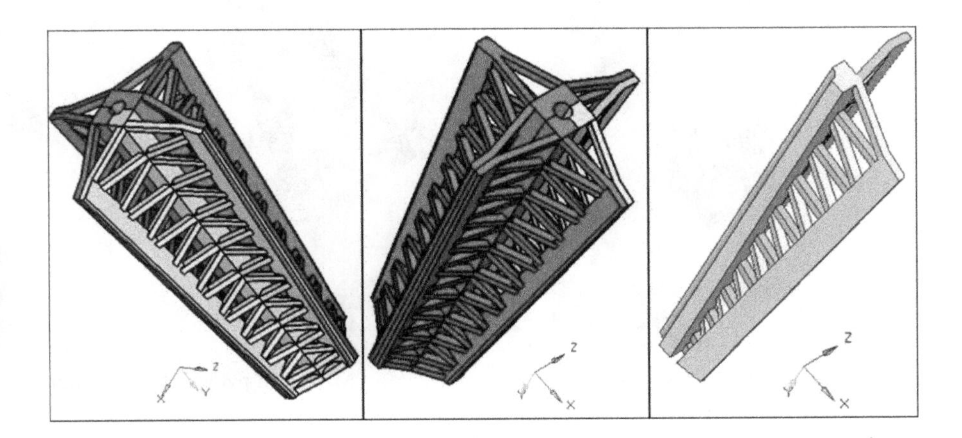

FIGURE 7.6 Geometric model of a pylon considered for FEA.

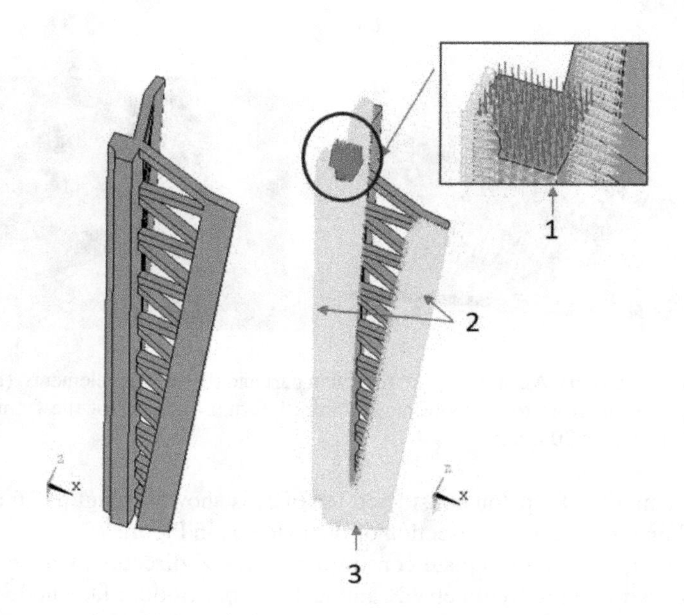

FIGURE 7.7 Loads applied on the square cross section of a pylon.

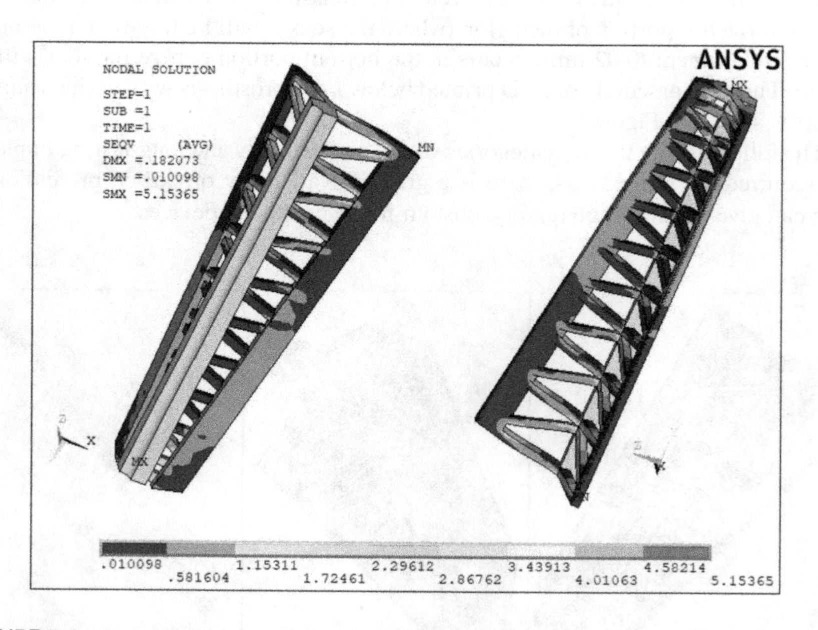

FIGURE 7.8 Von Mises stress distribution (values are in MPa) in a pylon. The maximum yield stress is 5.15 MPa.

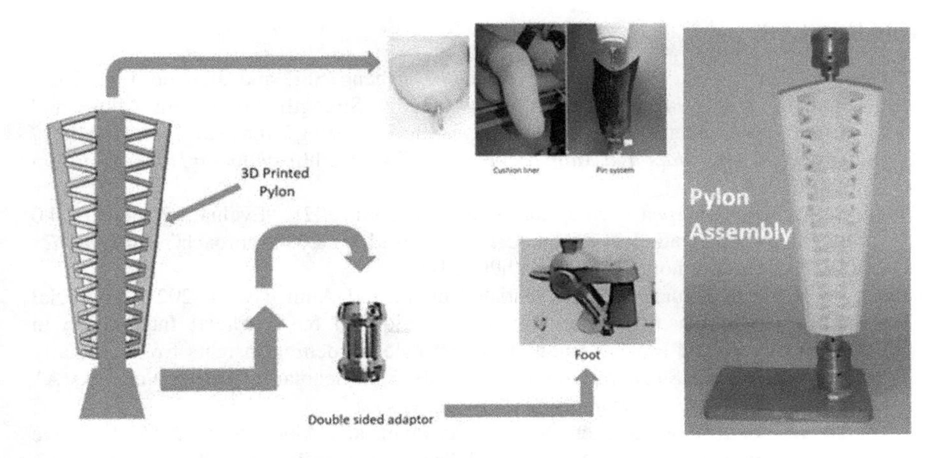

FIGURE 7.9 The final assembly of a 3D printed below knee prosthesis with various attachments.

Due to the enormous pressure exerted by a pylon, amputees who have been fitted with transtibial prostheses have experienced severe discomfort and pain. Furthermore, transtibial and lower limb prostheses are costly, heavy, time-consuming, and do not allow for natural walking action. The pylon's design incorporates trusses as the primary weight and stress bearing element, allowing it to withstand various forces and stresses induced during walking by compression of the truss members, allowing the amputee to walk naturally and providing a happier and easier prosthetic experience.

The practice of additive manufacturing to create custom-built lower limb prosthetic devices is being investigated. The ability to control material composition and manufacturing procedures is one of the advantages of additive manufacturing. Further testing will be required using computer models and 3D scanning in conjunction with 3D printing to create custom-made lower limb prosthetic devices.

7.10 CONCLUSIONS

In today's Industry 4.0, new software and digital updates are frequently released. Due to their flexibility to grow with the latest technology, 3D printing services remain vital in industrial processes. 3D printers will become progressively more important, particularly for rapid prototyping. The medical and dental fields benefit from 3D printing services. Medical teams can receive patient-specific products ranging from prostheses to implants to bioprinting. During the Covid-19 pandemic, 3D printing was used to produce ventilator valves and hospital hand-free door handles. 3D printing, also known as additive manufacturing, is a production necessity as part of the Industry 4.0 revolution. Key businesses and sectors require high quality complicated parts that can be produced in the least amount of time. As a critical tool in the quick prototype process, 3D printing satisfies these requirements.

REFERENCES

Altıparmak, Sadettin C., Victoria A. Yardley, Zhusheng Shi, and Jianguo Lin. 2021. "Challenges in Additive Manufacturing of High-Strength Aluminium Alloys and Current Developments in Hybrid Additive Manufacturing". *International Journal of Lightweight Materials and Manufacture* 4(2): 246–61. https://doi.org/10.1016/j.ijlmm.2020.12.004.

Chang, Shih-Chia, Hsu-Hwa Chang, and Ming-Tsang Lu. 2021. "Evaluating Industry 4.0 Technology Application in SMEs: Using a Hybrid MCDM Approach". *Mathematics* 9(4): 414. https://doi.org/10.3390/math9040414.

Jadhav, Priya, Arun Kumar Bongale, Satish Kumar, and Amit Tiwari. 2021. "Artificial Intelligence in Plasma Electrolytic Micro-Oxidation for Artificial Intelligence in Plasma Electrolytic Micro-Oxidation for Surface Hardening-Insights from Scholarly Citation Networks Surface Hardening-Insights from Scholarly Citation Networks A". *Library Philosophy and Practice (e-Journal).*

Jin, Yu-an, Jeff Plott, Roland Chen, Jeffrey Wensman, and Albert Shih. 2015. "Additive Manufacturing of Custom Orthoses and Prostheses: A Review". *Procedia Cirp* 36: 199–204.

Kumar, Sandeep, Monika Nehra, Deepak Kedia, Neeraj Dilbaghi, K. Tankeshwar, and Ki Hyun Kim. 2020. "Nanotechnology-Based Biomaterials for Orthopaedic Applications: Recent Advances and Future Prospects". *Materials Science and Engineering C* 106: 110154. https://doi.org/10.1016/j.msec.2019.110154.

Ngo, Tuan D., Alireza Kashani, Gabriele Imbalzano, Kate T.Q. Nguyen, and David Hui. 2018. "Additive Manufacturing (3D printing): A Review of Materials, Methods, Applications and Challenges". *Composites Part B: Engineering* 143: 172–196.

Paital, Sameer R., and Narendra B. Dahotre. 2009. "Calcium Phosphate Coatings for Bio-Implant Applications: Materials, Performance Factors, and Methodologies". *Materials Science and Engineering R: Reports* 66 (1): 1–70. https://doi.org/10.1016/j.mser.2009.05.001.

Pirouzi, Gh, N.A. Abu Osman, A. Eshraghi, S. Ali, H. Gholizadeh, and W.A.B. Wan Abas.2014. "Review of the Socket Design and Interface Pressure Measurement for Transtibial Prosthesis". *The Scientific World Journal* 2014.

Radosh, Aleksandra, Wiesław Kuczko, Radosław Wichniarek, and Filip Górski. 2017. "Prototyping of Cosmetic Prosthesis of Upper Limb Using Additive Manufacturing Technologies". *Advances in Science and Technology Research Journal* 11(3): 102–108. https://doi.org/10.12913/22998624/70995.

Rajťúková, Va, M. Michalíková, L. Bednarčíková, A. Balogová, and J. Živčák. 2014. "Biomechanics of Lower Limb Prostheses". *Procedia Engineering* 96: 382–391.

Song, Mi Hyun, Won Joon Yoo, Tae Joon Cho, Yong Koo Park, Wang Jae Lee, and In Ho Choi. 2019. "In Vivo Response of Growth Plate to Biodegradable Mg-Ca-Zn Alloys Depending on the Surface Modification". *International Journal of Molecular Sciences* 20(15). https://doi.org/10.3390/ijms20153761.

Soud, Wafa A., and Mohammed R. Ahmed. 2019. "Experimental Study of 3D Printing Density Effect on the Mechanical Properties of the Carbon-Fiber and Polylactic Acid Specimens". *Engineering and Technology Journal* 37(4A): 128–132.

Tavangarian, Fariborz, and Camila Proano. 2019. "The Need to Fabricate Lower Limb Prosthetic Devices by Additive Manufacturing". *Biomedical Journal of Scientific & Technical Research* 15(5): 11662–11673.

Wu, Y., R.D. Keagy, H.J. Krick, J.S. Stratigos, and H.B. Betts. 1979. "An Innovative Removable Rigid Dressing Technique for Below-the-Knee Amputation". *JBJS* 61(5): 724–729.

Yang, Wen Jing, Dicky Pranantyo, Koon Gee Neoh, En Tang Kang, Serena Lay Ming Teo, and Daniel Rittschof. 2012. "Layer-by-Layer Click Deposition of Functional Polymer Coatings for Combating Marine Biofouling". *Biomacromolecules* 13 (9): 2769–2780. https://doi.org/10.1021/bm300757e

8 Technology Gap Analysis with Respect to Mysore Printing Cluster

An Attractive Opportunity in Industry 4.0 Market

G. Devakumar

CONTENTS

8.1 INTRODUCTION

8.1.1 MYSORE PRINTERS CLUSTER

Mysore Printers Cluster is a group of micro, small and medium enterprises involved in printing, chiefly using various traditional methods such as screen-printing technology, offset printing machinery, single colour printing, and drum-type and

DOI: 10.1201/9781003200857-8

plate-type press-based impression printing methods. There are about 250 enterprises in this cluster. Some of the major products of the Mysore Printers Cluster are business cards, text books, pamphlets, journals, wedding cards, labels, letter heads, hotel menus, calendars, receipt books, and ID cards.

8.1.2 Aim

The main aim of the Technology Information Forecasting and Assessment Council (TIFAC)-funded project is to study the present technological and marketing practices of the Mysore Printers Cluster, identify the technological gaps, and provide suitable recommendations for the sustainability of the enterprises, with a view to creating an attractive opportunity in the Industry 4.0 market.

8.1.3 Objectives

The broad objectives of the study are

- To carry out a literature survey related to printing technologies in traditional printing units and advanced printing industries across the globe.
- To carry out a preliminary background study on existing printing technology, the number of industries involved, and the available variety of products.
- To conduct a pilot study and prepare a survey questionnaire from primary and secondary data.
- To carry out a survey to understand the educational qualifications of people involved in the present status of the printing technologies and the turnover of the printing cluster in and around Mysore, Karnataka.
- To identify the printing materials covering the entire production and value chain and to make a comparison of these technologies with advanced technologies.
- To identify the existing technological gaps with respect to energy, the environment, and productivity, as well as in the manufacturing processes in the cluster to ensure progressive growth for the sustainability of potential entrepreneurs.
- To highlight the causes of the existing technological gaps, suggest remedial measures and suitable machines/suppliers with their cost, and present them in the form of a report with a projected market.
- To recommend a plan for making focused technological interventions in the cluster with a view to creating an attractive opportunity in the Industry 4.0 market.

8.1.4 Methodology

1. A literature survey was carried out referring to national and international journals related to printing technologies and machinery, books,

publications, magazines, and also by accessing various scholarly websites to obtain secondary data.
2. Key personnel in the existing printing cluster were visited and a pilot study was conducted to obtain the various inputs and challenges faced by them, in order to formulate the questionnaire. Also, interactions took place with various local government organisations and industrial associations in order to obtain key insights about the printing cluster.
3. Upon validation of the final questionnaire, a survey was carried out by meeting the individual key personnel from all the printing units in and around Hebbal Industrial Area, Mysore.
4. The current materials and production process used by the cluster members were analysed by means of a method study.
5. Gap analysis was carried out based on the primary data collected from the questionnaire, and also from the secondary data.
6. A project report was prepared and submitted, highlighting the cause of the existing technological gaps and suggesting remedial measures for the future sustainability of the printing cluster. An awareness workshop was conducted for the cluster members to educate them and to help establish better coordination with the cluster actors.

8.1.5 Vision

Keeping the stiff competition in mind, the current cluster members are encouraged to find ways to provide superior customer satisfaction with an on-time delivery strategy by means of employee motivation and a reduction in wastage and input costs, and to strive for export opportunities with fair wage practices.

8.2 MYSORE CITY

Mysore city is located in the southern part of the Indian state of Karnataka. It is the third largest city in the state and is located about 146 km southwest of Bengaluru. It is approximately 128.42 square kilometres in area and has a population of 920,550 according to the 2011 Census of India [1].

8.3 PRINTING TECHNOLOGIES

8.3.1 Introduction to Different Printing Technologies

Printing is well known for its reproduction process in which printing ink is applied to a printing substrate in order to transmit information (graphics, images, text) in a repeatable form using an image-carrying medium (e.g. a printing plate).

The main history of printing goes back to the replication of depictions/pictures by means of stamps during the early times. The main use of round seals for rolling an impression into clay tablets goes back to the early Mesopotamian civilisation, some 3000 BCE. These are the most common works of art to survive, and feature complex and beautiful images. In China and Egypt, the use of small insignificant stamps as seals preceded the use of larger blocks. In China, India, and Europe, cloth printing preceded paper/papyrus printing, with the processes essentially the same. In Europe, special presentations of impressions were printed on silk, until the 17th century. The development of printing made it possible for newspapers, books, and other reading material to be produced in countless numbers, and it played an important role in encouraging literacy. It is surmised that Johannes Gutenberg, from the German city of Mainz, developed European moveable type printing technology with the printing press around 1439, and in just over a decade, the European age of printing began.

Printing technology can be broadly classified as conventional and non-impact printing, as shown in Figure 8.1. The conventional printing technology uses master plates. Before starting the actual printing process, a master plate is prepared. The master plate contains the image to be printed onto a substrate such as paper or polyethylene. Depending on how the master plate is prepared, there are various printing technologies.

Considering the conventional printing technology (which utilises master plates), technically speaking, there are four fundamental methods of printing:

- Letterpress printing (image and non-image areas are in different planes)
- Gravure printing (image and non-image areas are in different planes)
- Planographic printing (image and non-image areas are in the same plane)
- Screen-printing (technique based on screen and stencil)

a. **Letterpress printing:** The image areas (which are coated with a layer of ink by means of rollers) are raised above the non-image areas. Ink is subsequently transferred onto the desired substrate.
b. **Gravure printing:** The image elements are inscribed into the surface of a cylinder. The non-image areas are at a constant original level. Prior to printing, the complete printing plate (non-printing and printing elements) is swamped with ink. Ink is removed from the non-image areas (by a wiper or blade) before printing, so that the ink remains only in the cells. The ink is transferred from the cells to the printing substrate by high printing pressure and the adhesive forces between the printing substrate and the ink.
c. **Planographic printing:** The printing and non-printing parts are on the same level. A few examples are lithography, offset, and collotype printing.
   Collotype printing was the state of the art for photographic imitation at the turn of the 20th century. However, due to the high level of expertise required to obtain reliable results, it was quickly substituted by the faster, inexpensive, and more mechanised process of offset printing [6]. As regards lithography/offset, the distinctive feature of the printing areas is the fact that they are ink-accepting, whereas the non-printing plate fundamentals are ink-repellent. This effect is shaped by physical, interfacial surface phenomena [3].

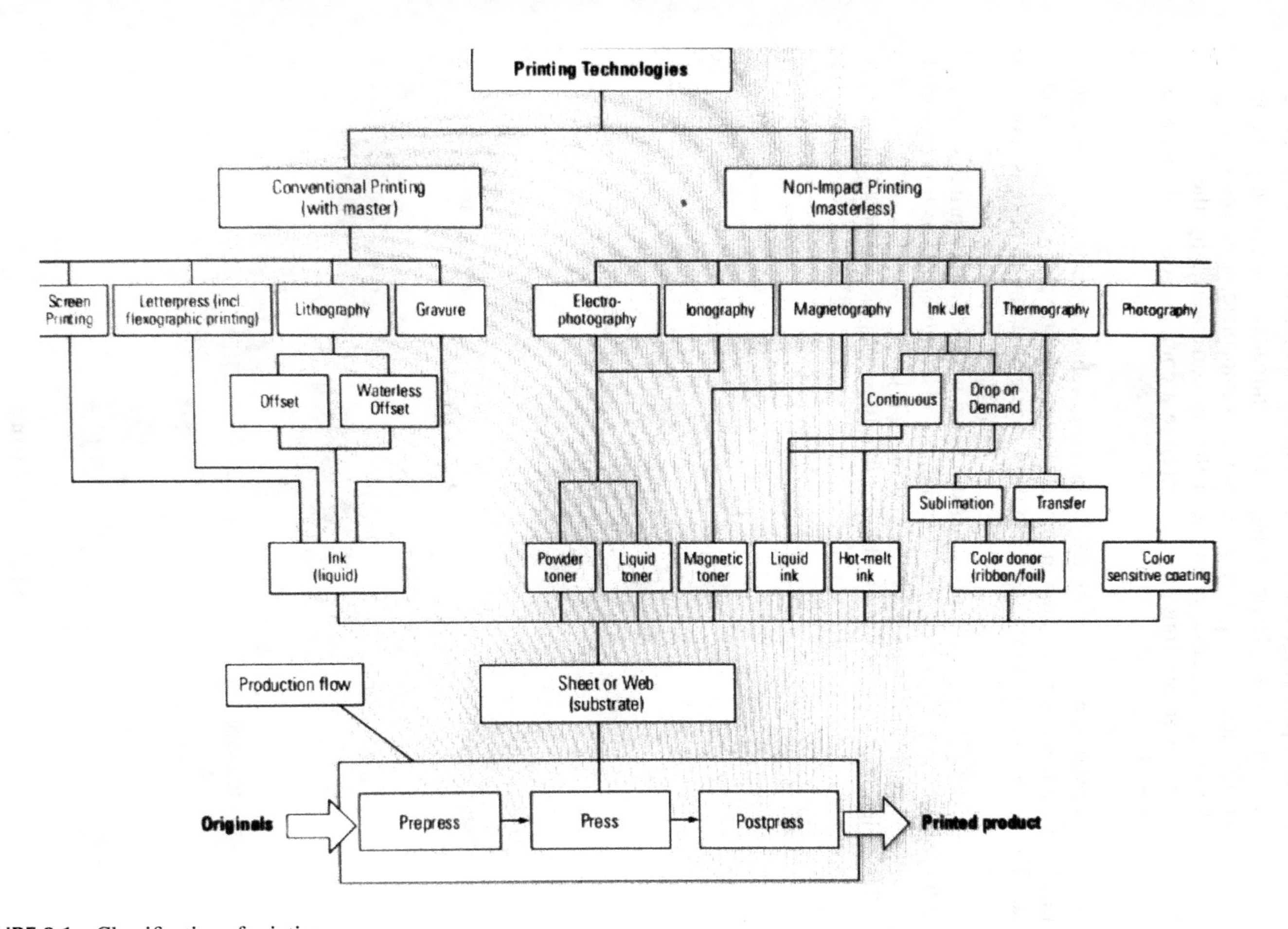

FIGURE 8.1 Classification of printing.

d. **Screen-printing:** Ink is forced through a screen (fine fabric made of natural silk, plastic, or metal fibres/threads). The screen-printing stencil helps as a printing plate. Ink is imprinted or moved through the image-specific, open mesh that is not covered by the stencil. The screen-printing plate is therefore a combination of screen and stencil.

The fundamental principles of these four printing technologies are succinctly illustrated in Figure 8.2, which shows the image carriers (master plate) in all four methods. The classification of printing technology requiring a master plate is shown in Figure 8.3.

In this chapter, the main technologies discussed will be gravure printing, flexography (which comes under letterpress printing), offset printing (which is classified under planographic printing), and screen-printing [7].

The printing process is usually divided into three phases: pre-press, press, and post-press.

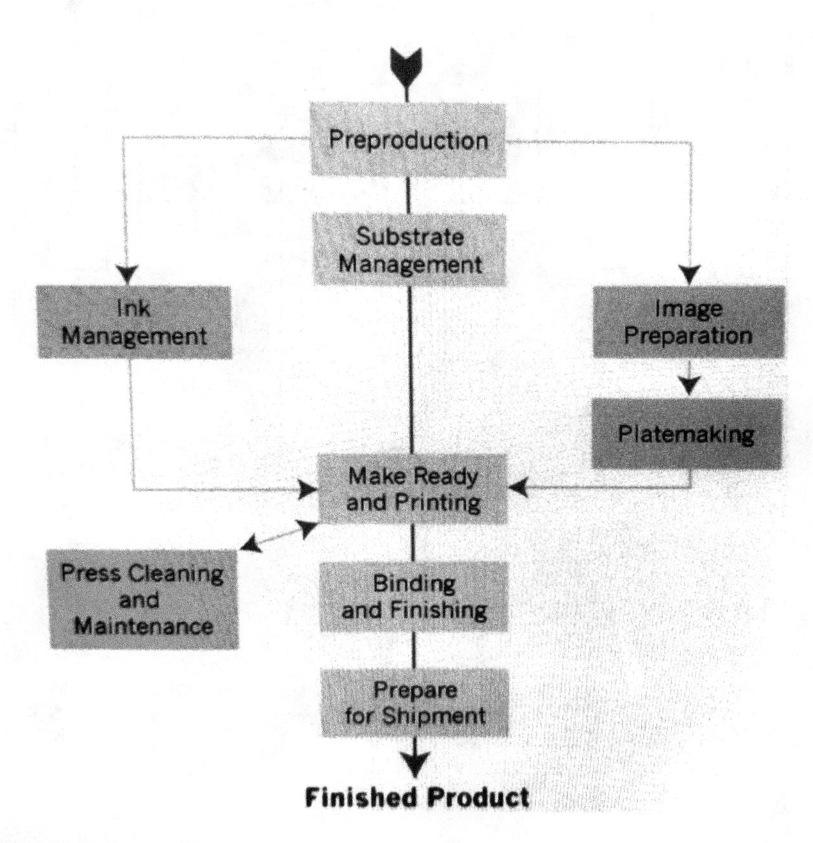

FIGURE 8.2 Finished product flow process chart.

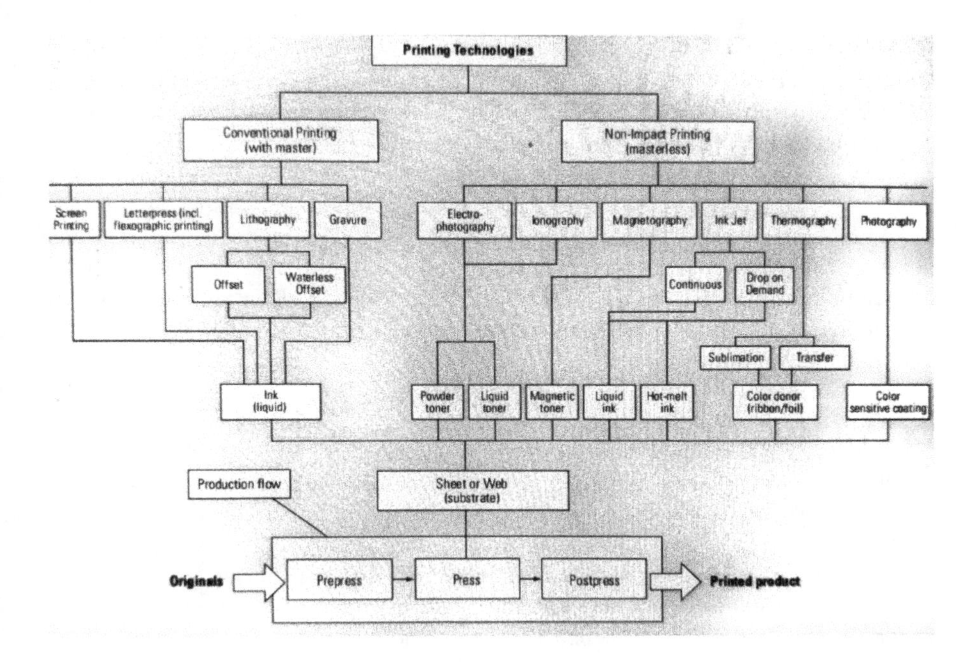

FIGURE 8.3 Flexography.

8.3.1.1 Process Flow of Flexography

The main operations involved in flexography printing are plate-making, image preparation, printing, and finishing. The process of printing each colour on a flexo press consists of a series of four rollers:

- Meter roller
- Ink roller
- Plate cylinder
- Impression cylinder

The ink is primarily moved from an ink pan to the meter roller or anilox roller (which is the second roller) by the first roller. The anilox roller then meters the ink to an unchanging thickness onto the plate cylinder. The substrate then moves between the plate cylinder and the impression cylinder, which is the fourth roller [10].

The impression cylinder puts pressure on the plate cylinder, thereby moving the image onto the substrate. The web, which has now been printed, is fed into the overhead dryer so that the ink is dried before it enters the next printing unit [4].

After the substrate has been printed with all colours, the web is fed through a supplementary overhead tunnel dryer to remove most of the residual solvents or water. The complete product is then rewound onto a roll or is nursed through a cutter [5].

8.3.1.2 Pre-Press, Plate-Making, Handling, and Storage

The photochemical pre-press process of flexographic photopolymer printing plates comprises a number of stages when forming: back exposure; main exposure; washing, drying, and stabilisation; post-exposure; and light-finishing.

Many have pointed out that consistency in plate-making is the key point in establishing a proper printing plate for good print quality. Handling, cleaning, and storage are of great importance in ensuring and maintaining good print quality. Storage and handling conditions have been investigated, and positioning and exposure to light, and temperature are factors that need to be controlled. In cleaning, knowledge of solvents is of great importance as they can produce properties such as shrinkage, cracking, and hardening [8].

8.4 TECHNOLOGY GAPS

A technology gap analysis deals with the analysis of the variance between the existing state (as is) and the desired state (to be) in the future. It is likely to find the real level of knowledge in an organisation by analysing present technology gaps in the organisation and comparing them with established standards in this area. Updating to advanced technology may require additional investments for a proprietor/owner/partner. Hence, the gap that needs to be overcome may involve crossing a barrier, called cost.

Table 8.1 identifies such gaps. The advantages accrued and other details are discussed in detail in Chapter 9. The gap in the technology covers offset printing, flexography, and gravure printing. However, while conducting a preliminary survey, we noticed that offset technology is a ubiquitously preferred technology for various reasons such as relevance, printers' area of interest, application, ease of use, cost, preparation of make-ready, among other factors. Hence, the focus of our study was more on technology based on offset printing.

In Table 8.1, the column titled "Indicator" indicates whether the technology mentioned in the "Advanced" column results in

- Decreasing energy consumption (shortened as **Eng**)
- Decreasing impact on the environment (shortened as **Env**)
- Increasing productivity (shortened as Prd)

8.5 SUGGESTED REMEDIAL MEASURES AND RECOMMENDATIONS WITH A VIEW TO CREATING AN ATTRACTIVE OPPORTUNITY IN INDUSTRY 4.0 MARKET

In Chapter 7, the technologies used by traditional printers and global industries were compared and some gaps were identified and listed. In this chapter, the points made in Chapter 7 have been elaborated and their advantages stated.

Below is a list of recommendations with regard to various stages of printing in different printing technologies. The advantages of the suggestions are also provided.

TABLE 8.1
List of Identified Gaps

Area of technology	Traditional	Advanced	Indicator
Offset printing	The master plates are exposed and developed manually.	The main use of computer-to-plate (CtP) technology in pre-press for offset printing.	**Eng, Env, Prd**
	Using metal halide lamps for plate exposures.	Use of UV LED lamps instead for better energy efficiency.	**Eng**
	Plate-making using a developer solution.	Thermal plates requiring no chemistry.	**Eng, Env, Prd**
	Use of mineral oil–based inks.	Use of vegetable oil–created inks.	**Env**
	All of the surveyed offset technology utilises an IPA-based dampening solution to separate the image area from the non-image area.	Waterless offset eliminates the use of IPA-based dampening solution, VOCs, etc.	**Env**
	Lack of any means to quantitatively monitor the emissions of VOCs.	Accurate IPA measuring instruments (e.g. measuring the IR absorption spectra in the gaseous phase or measurement of the acoustic velocity) are commercially available.	**Env**
	No dedicated team to actively monitor the amount of resources wasted, such as paper and water.	Employment of a dedicated monitoring team.	**Env, Prd**
	Alignment of films for subsequent exposure; preparation of make-ready, washing, etc., was done manually.	Automatic wash-up devices, automatic plate changing, and automatic alteration of paper guiding elements are in general use.	**Prd**
	To fill the ink fountains on sheet-fed offset presses with ink, a spatula is used directly from the ink can.	Ink cartridges that are simpler and more convenient for filling up the ink fountains on sheet-fed presses are used [2].	**Prd**
	Plate removal and feeding operations are performed successively for individual plates with the assistance of the operator.	This can be achieved automatically by means of cassette systems with no operator interference once the change-over process has been initiated [2].	**Prd**

(*Continued*)

TABLE 8.1 (CONTINUED)
List of Identified Gaps

Area of technology	Traditional	Advanced	Indicator
	Manual sheet size and paper travel adjustment: Positioning of the paper pile in the feeder and setting the suckers to pick up the sheet and transport it into the first printing unit, sheet guidance and alignment elements on the feeding table are finished physically.	With present-day technologies, remote adjustment and position measurement procedures can be automated. This is possible for both sheet feeders and sheet deliveries.	**Prd**
	Silver halide solutions may not be recycled.	The service life of the developing bath can be stretched by developer-saving systems and long-life chemicals. The recycling of recovered silver is also possible for films and photographic papers with a blackening factor of more than 30% [2].	**Env**
	Sufficient measures are not taken towards eliminating VOCs (i.e. conventional offset printing using mineral oil–based ink and IPA-based dampening solutions). Typically, 8–15% of IPA is added to dampening solutions.	VOC emissions can be reduced by using less alcohol content, which can be achieved by using suitable substitutes such as glycol. Other technical improvements are the use of hydrophilic dampening rollers, reverse osmosis water processing, inking system temperature control, etc., helping to reduce IPA by 4–8% or even less [2].	**Env**
	Significant amounts of VOCs are consumed in heatset web offset printers	Use of UV inks greatly decreases VOC emissions.	**Env**
Flexography	Plate-making is done physically by means of UV exposure and washing; the side walls of the printing elements cannot be shaped if laser engraving is used.	Using laser engraving to plate cylinders that have been fully coated with elastomer and then precision-ground, superior quality is obtained; the problem of dot gain can be overcome by employing direct laser engraving.	**Env, Prd**
	A work cycle that presents problems as regards process reliability, when plates are chemically developed and cleaned.	Computer-to-plate (CtP) leads to a marked increase in printing quality. Other advantages are lower dot gain and a greater range of print contrast.	**Env, Prd**

(*Continued*)

TABLE 8.1 (CONTINUED)
List of Identified Gaps

Area of technology	Traditional	Advanced	Indicator
	Analogue sheet photopolymer: Plate-making with solid photopolymer sheets is done using photographic (silver halide) negative films.	Digital sheet photopolymer: Rather than using a photographic negative film to deliver the image mask, the digital mask layer is selectively ablated away using an IR laser to create an integral mask; this eliminates the use of a negative film and provides a streamlined and fully digital workflow, and improves image quality as well.	**Env, Prd**
	Solvent plate processing: This method has been used for years; after UV exposure of the photopolymer plate to create an image, the non-image areas that have not been exposed are removed by means of a solvent-based method.	Thermal plate processing: No solvent is used in this process, rather the unpolymerised photopolymer that is leftover after UV exposure is removed by melting and then wicking away using a blotter material; this is environmentally friendlier.	**Env, Prd**
	Solid photopolymer: This method involves buying a solid plate (sheet), ready for imaging (either by analogue, or now more commonly, digital means).	Liquid photopolymer: Liquid photopolymer is cast by the platemaker at the time of imaging; typically less costly, quicker to produce, and generates little hazardous waste.	**Env, Prd**
	Inks based on alcohol: Solvent-containing (VOC-containing) flexographic printing [8,9] inks contain mainly alcohol (ethanol) and often ethyl acetate, special grades of petroleum spirit, toluene, IPA, and acetone.	Water-based inks and UV inks: Water-based (VOC-free) flexographic printing inks predominate in printing on absorbent materials; VOC-free UV flexographic printing inks are also used [2].	**Env**
Gravure	Rigid gravure cylinders are used.	Gravure wrap-around plates can be used for quick maintenance.	**Prd**
	Cylinder preparation is carried out by etching.	Laser and mechanical-based stylus system yields better quality and is environmentally friendlier.	**Env**
	Toluene-based inks are used.	Solvent-free ink developed by Hottech Company.	**Env**

8.5.1 Action Plan

Based on a literature survey, primary data, secondary data, recommendations, and suggestions provided for this project, an action plan is provided in Table 8.2, emphasising various training modules with topics, and a brief content is proposed for future training requirements at the Mysore Printers Cluster to enhance the knowledge and cognitive skills among the various levels of employees for market sustainability. This can be carried out by employing domain experts to conduct workshops at the grassroots level.

8.6 CONCLUSION

In order to enhance the sustainability of the Mysore Printers Cluster with a view to creating an attractive opportunity in the Industry 4.0 market, it is suggested to adopt a new approach that will improve its prospects in the future. The cluster confronts certain key issues such as lack of loan facilities at reasonable interest rates, attrition as well as a lack of skilled workers, raw material accessibility, obsolete machinery, and lack of a marketing team. These issues were highlighted in discussions with the cluster members, and while conducting the survey. The findings of this study are summarised as follows:

- Offset technology is predominantly used in the Mysore Printers Cluster because it is the most suitable and economical printing technology for the demands that they primarily address, which are printing products such as text books, journals, and pamphlets. It is observed that more than 70% of the cluster members are equipped with offset printing technology.
- From the survey findings, it is observed that the majority of the cluster members are higher secondary certificate (HSC) qualified, followed by undergraduate (UG), diploma, and Industrial Training Institute (ITI) qualified.
- It is found that the majority of the cluster members are expecting growth by up to 10% in the coming years, while about 5% are expecting a constant or decreased turnover.
- The terms "energy efficiency", "environmental impact", and "productivity" are important indicators in a printing firm. A few parameters are shortlisted under each of the categories: energy, environment, and productivity. Based on the measures being taken by an enterprise in contributing to each of the defined categories, it came to light that the majority of them fall in the range of 3, on a scale of 1–5.
- Studies conducted by various agencies/companies such as McDermid Printing Solutions, Minnesota Technical Assistance Programme, Printers' National Environmental Assistance Centre, and Best Available Techniques, their suggestions pertaining to improving energy efficiency, reducing environmental impact, and increasing productivity are recommended in this study. Some are related to heatset and non-heatset offset printing,

TABLE 8.2
Action Plan

Field	Area	Content
Management	Sales and marketing:	• Customer need and want analysis • Supply vs demand • Customer satisfaction
	Customer acquisition:	• Identifies and converts prospective customer • Champions for the organisation • Creates customer experience • Maintains positive long-term relationship • Leverages positive experiences to create customer loyalty
	Creative thinking:	• Innovativeness • Incorporates existing ideas and novel ideas • Unique approach to resolve problems and capitalise on opportunities • Professionalism
Technical	Environment:	• Thermal plates in offset printing and its role in environmental aspects • Use of vegetable oil–based inks • Waterless offset and its advantages pertaining to environmental aspects • Isopropyl alcohol (IPA)-measuring instrument • Wastage monitoring team • Recycling of recovered silver in offset printing • Use of glycol instead of alcohol to curb emissions of volatile organic compounds (VOC) • Use of UV inks • Laser engraving plate cylinders in flexography printing • Computer-to-plate technology in flexography • Use of a digital sheet polymer in flexography • Thermal plate processing of a photopolymer plate • Liquid photopolymer in flexography • Water-based and UV-based inks in flexography • Cylinder preparation using laser and mechanical-based stylus in gravure printing • Solvent-free ink in gravure printing
	Energy:	• Use of computer-to-plate technology and its role in energy aspects • Use of UV LED lamps and their role in energy aspects • Thermal plates in offset printing and their role in energy aspects

and the environmental impacts of different plate-making technologies in flexography.

- It is noticed that emissions of harmful volatile organic compounds (VOCs) have been a major concern. The major sources of VOCs are printing inks and dampening solutions. Across the globe, endeavours have been made to address this issue, and some of the commercially available alternatives have been reported in this study, e.g. vegetable oil–based printing inks, UV inks, and waterless offset printing.
- In this study, recommendations are made to upgrade the machinery in various stages of printing, such as pre-press, press, and post-press.
- It is suggested to establish a common facility centre to address some of the issues mentioned above. The common facility centre would primarily aid in curbing the losses suffered due to the lack of advanced machinery and any other facilities.

ACKNOWLEDGEMENT

We would like to sincerely thank the Technology Information Forecasting and Assessment Council (TIFAC) New Delhi for choosing Ramaiah University of Applied Sciences (RUAS) as knowledge partner of this sponsored project, and we are deeply indebted to TIFAC for giving us this opportunity.

We humbly express our gratitude to Shri. Sureshkumar Jain, General Secretary of Mysore Industries Association, and his colleagues for arranging various discussions/meetings with the cluster members.

I wish to thank the team members of Techno Centre for their support and also thank all the members of Mysore Printers Cluster for giving honest feedback and for their whole-hearted and active support, and Techno Centre for completion of this project.

REFERENCES

1. Kipphan, H., 2001. *Handbook of Print Media*. Berlin: Springer.
2. Wikipedia, 2015. History of Printing. [online] Available at: http://en.wikipedia.org/wiki/History_of_printing.
3. Benrido, Contemporary Collotype, Hariban Award, Collotype Photo Competition, 2015. Home: Benrido, Contemporary Collotype, Hariban Award, Collotype Photo Competition. [online] Available at: https://benrido-collotype.today.
4. Coatings.specialchem.com, 2015. Overview Gravure Printing Technology, Market Share, Growth...Few figures – SpecialChem. [online] Available at: http://coatings.specialchem.com/editorial/overview-gravure-printing-technology-market-share-growth-few-figures.
5. Pneac.org, 2015. Printing Process Descriptions: Environment and Printing: The Printers' National Environmental Assistance Centre: PNEAC: The Environmental Information Website for the Printing Industry. [online] Available at: www.pneac.org/printprocesses/gravure/sheetfed.cfm.

6. Era.eu.org, 2015. European Rotogravure Association. [online] Available at: www.era.eu.org/04/advantages.html.
7. The Printing Process: Offset Printing, 2012. Available at: https://ohsobeautifulpaper.com/2012/01/the-printing-process-offset-printing/.
8. Johnson, Johanna, 2008. Aspects of Flexographic Print Quality and Relationship to Some Printing Parameters, Karlstad University Studies, ISSN 1403–8099.
9. WhatTheyThink, 2015. Flexographic Printing: Five Critical Steps Forward. [online] Available at: http://whattheythink.com/articles/71516-flexographic-printing-innovation.
10. Nclarion.weebly.com, 2015. PRINTER: Advantage and Disadvantage of Inkjet Printers. [online] Available at: http://nclarion.weebly.com/advantage-and-disadvantage-of-inkjet-printers.html.

9 Intelligent Machining

Ajit Dhanawade, Seema Wazarkar, and Vishal Naranje

CONTENTS

9.1 INTRODUCTION

The machining process is characterised by numerous process parameters viz. cutting speed, feed, depth of cut, tool geometry, and tool condition. Therefore, machining processes are fundamentally intricate, non-linear, multifaceted, and subject to uncontrollable factors (Astakhov and Outeiro, 2008). Machining processes are performed by skilled machine operators. During machining, decision-making is a very important task. The machine operator makes decisions on the basis of instinct and experience. It is very difficult to develop a fully automatic machining system that can perform machining-related activities like a skilled machine operator. To develop a fully automatic intelligent machining system, the machining system requires the capabilities to control, monitor, and diagnose the process. An intelligent machining system directly controls the machining process by adopting models based on machining and real-time sensor data (Zhou et al., 2018). The modelling of a machining process is necessary for building up an intelligent machining system. The model of a machining process depends on the physics of the process. The model is developed in mathematical form, mostly in terms of differential equations. The finite element analysis method (FEM) can be used to solve the differential equations. In some cases, the mathematical model is developed on the basis of data received from the process, which describes the behaviour of the process. An intelligent machining

DOI: 10.1201/9781003200857-9

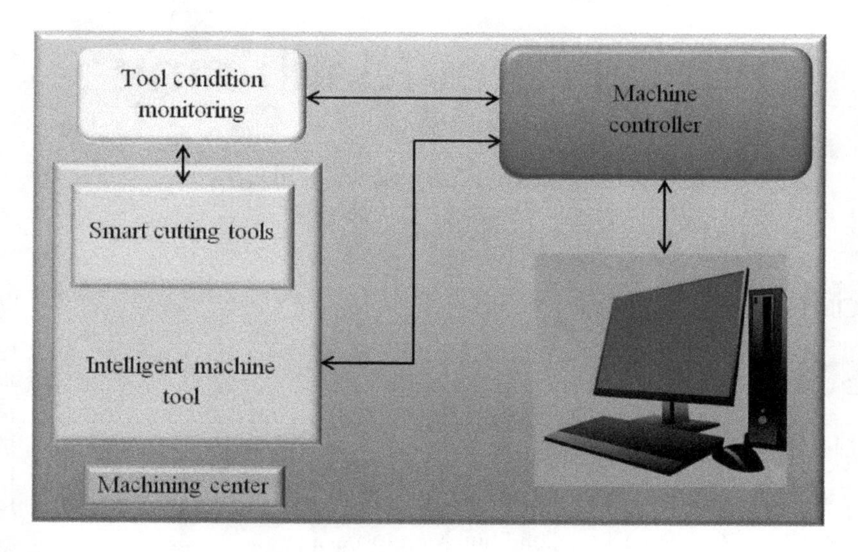

FIGURE 9.1 Intelligent machining system.

system can be developed through building up an intelligent machine tool and cutting tools, and intelligent techniques for planning, control, and decision-making. Figure 9.1 shows a typical intelligent machining system.

Intelligent machine tools can judge the machining process and take the optimum decisions. In the process of adapting to a complete automation of machine tools, the most difficult part is to replace labour with intelligence. Therefore, an intelligent machine tool automatically monitors and analyses the data of the machining process, its surroundings, and the data itself. The intelligent machine tool achieves optimal behaviour under uncertain surroundings and their changes (Chen et al., 2019).

Smart cutting tools and real-time condition monitoring of tools are essential in achieving the desired industrial advantage in terms of low cost and high quality products, productivity improvement, and damage prevention. Smart cutting tools are useful in a high precision machining process for proactive machining to handle the complexity, dynamic behaviour, and variations of the process (Cheng et al., 2017). Real-time tool condition monitoring systems accurately detect the growth of tool wear during machining operation. The system predicts abrupt failure of the tool before damaging the component under machining.

Soft computing techniques are useful in solving machining problems related to process variables. Machining processes are largely governed by process variables; therefore, finding the optimum combination of process parameters is an essential task to avoid the noise factor. Soft computing techniques such as fuzzy logic and nature-inspired algorithms are useful for optimisation. Optimising process parameters determine the approximate solutions to the indefinite physics of the process (Deb and Dixit, 2008).

9.2 MACHINE TOOLS

9.2.1 Evolution

Machine tool technologies are applying their efforts to bring about the complete automation of equipment, especially machines used in manufacturing industries. A major concern in the complete automation of machines is how to develop competitive intelligence to replace manual and intellectual labour. Industry 4.0 and corresponding initiatives in manufacturing industries aim at the integration of advanced information technology and artificial intelligence (AI) with intelligent manufacturing (Zhou et al., 2018). The development of traditional machine tools into intelligent machine tools follows three stages, i.e. numerical control machine tools, smart machine tools, and intelligent machine tools.

In numerical control machine tools, a numerically controlled system is used to transfer the manual labour to the controller. It acts between human and machine tools. The required information for machining is sent to the controller in terms of G-codes. This machine tool completes the machining by controlling the machine tool because maximum labour is shifted to the controller. Therefore, it is a vital component in machining operations. But numerical control is unable to sense, get feedback, or model the state of the machine tools while working, which may result in deviations in product quality and process efficiency. This means that machine tools lack intellectualisation. Figure 9.2 shows a numerically controlled machine tool along with a human cyber-physical structure. Globally, researchers have contributed their efforts to recent research on numerical controlled machine tools. Researchers

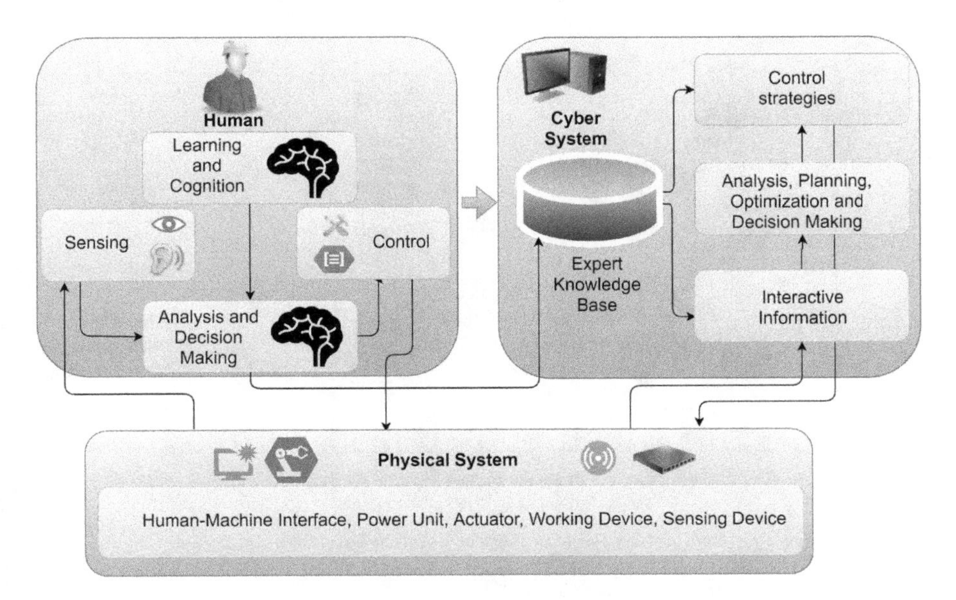

FIGURE 9.2 Numerically controlled machine tool with human cyber-physical structure.

have attempted to advance numerical controlled machine tools in terms of the optimal approach to multiple tool selection (Chen and Fu, 2011), humanity design (Jiang and Cheng, 2012), Therblig-based energy supply modelling (Lv et al., 2014), energy consumption (Lv et al., 2016), a mechanical model of errors of probes (Jankowski and Wozniak, 2016), an online machining error estimation method (Zhao et al., 2016), an identification approach to relative position relationships by laser tracker (Wang and Guo, 2019), a trajectory smoothing method using reinforcement learning, a kernel-based simulation (Schmid et al., 2020), a digital twin system (Wei et al., 2020, Shen et al., 2020, Pan et al., 2021), modelling for energy consumption (Pawanr et al., 2021), and networked numerical control systems using an instruction compression method (Liu et al., 2021) for numerical controlled machine tools to perform several operations.

Further integrated network and internet technologies are used in numerical controlled machine tools to develop smart machine tools. These machines use the internet along with sensors, making them capable of sense and interconnectedness. The sensors used in these machine tools sense the condition of the machining operation. The internet helps in collecting and aggregating the data of the machining operation. The real-time and non-real-time feedback system controls the device through analysing and processing the collected data. Therefore, numerical control machine tools have a specific level of intellectualisation. However, these machine tools are far from replacing manual labour and intellectual work with a competitive intelligence. Figure 9.3 depicts a smart machine tool along with a human cyber-physical structure. Globally, researchers have applied their efforts to developing smart machine tools; for example, a design methodology for smart actuator services (Desforges et al.,

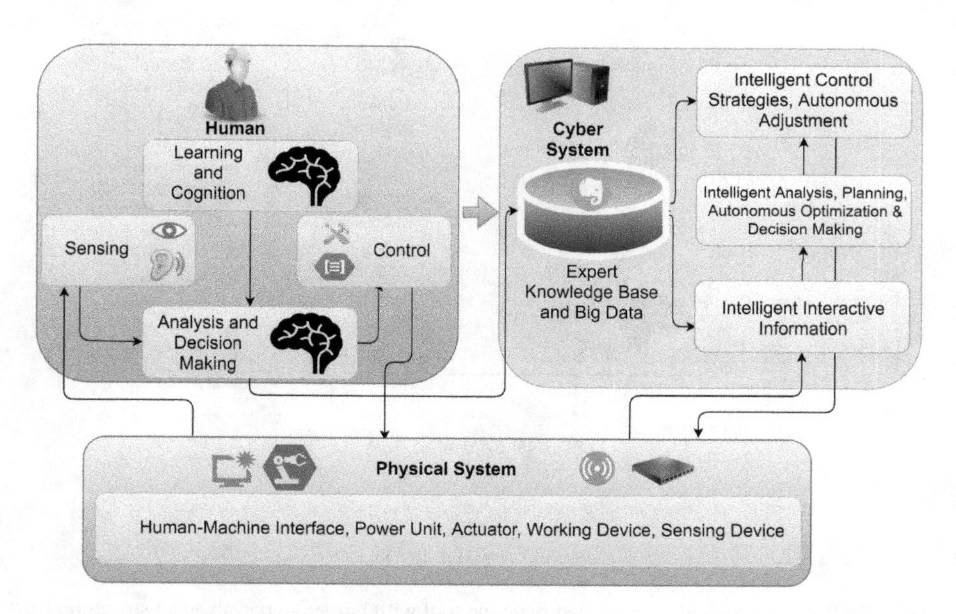

FIGURE 9.3 Smart machine tool and a human cyber-physical structure.

2011), modelling using smart interlocking software blocks (Nassehi and Newman, 2012), federate optimisation (Park et al., 2018), a modular smart controller (Barton et al., 2019), optimisation of production and maintenance activities (Gupta et al., 2019), smart retrofitting (Al-Maeeni et al., 2020), voice-enabled assistants (Longo et al., 2020), smart manufacturing multiplex (Botcha et al., 2020), a strategy for predictive maintenance (Costa et al., 2020), and a labelled Petri net–based approach for online fault diagnosis (Paiva et al., 2021).

9.2.2 New Generation AI

New generation AI depends on the growth of information technology. A prominent feature of this technology is to create, accumulate, and use the knowledge. A combination of AI and innovative manufacturing systems, i.e. intelligent manufacturing, is essential in Industry 4.0 (Chen et al., 2019). This AI technology in manufacturing offers an opportunity to develop intelligent machine tools with real intellectualisation. An intelligent machine tool can be a machine tool that automatically monitors and analyses the data of the machining process, its surroundings, and the data itself. This machine tool uses autonomous sensing and connection to retrieve the information about machining operations, working conditions, and surroundings. The machine tool controls the process and executes decisions through its key features, including generating data from learning modelling and optimising the decision-making cycle (Chen et al., 2019).

9.2.3 New Generation AI-Based Intelligent Machine Tool

An AI-based intelligent machine tool works on independent sensing connection, learning modelling, decision-making optimisation, and control execution. Researchers have made attempts to develop intelligent machine tools; for example, Ramesh et al. (2013) attempted the automation of machine tools using intelligent automation to fulfil the requirements of rapid computer-aided manufacturing. A real-time open architectural motion controller is used to realise the system in a three-axis vertical machining centre. O'Driscoll (2015) developed a non-invasive intelligent energy sensor to understand the state of the machine tool while working and to improve its efficiency and effectiveness. Li et al. (2017) developed an intelligent method for the optimum wheel position for designed grooves in the manufacture of milling and drilling tools. Vieler et al. (2017) proposed a standardised interface for an intelligent milling machine along with a cyber-physical system. It is useful in integrating the new tools in a machine without any mechanical change. Mahboubkhah et al. (2019) investigated the capability of an intelligent milling machine and its rotational ability around the horizontal axis. Durmaz and Yildiz (2019) studied the life of TiAlSiN-, AlCrN-, and TiAlN-coated carbide tools in intelligent machining. Liu et al. (2020) proposed a prognostic approach on the basis of adaptive variation mode decomposition. Further, a deep learning model has been developed to estimate tool life. Netzer et al. (2020) developed an approach to recognising online patterns based on numerical control codes. The numerical control codes are based on mean shift clustering matched with drive signals. Liu et al. (2020) proposed a framework based on bi-directional data and control flows. Intelligent monitoring of a machine tool and a data processing system is developed on the basis of an integrated model. Liu et al. (2021) developed an intelligent scheduling method based on a digital twin and super network. The method is useful for rapid and efficient process plan generation. Schmucker et al. (2021) developed an efficient intelligent system architecture that is useful for determining the tool tip dynamics in machining processes. The system enables the acquisition of in-process machine data and data from external sensors.

Figure 9.4 shows an intelligent machine tool system. The control unit of the intelligent machine tool is composed of a numerical controller, servo motor, servo drive, encoder and sensor, and so on. The control unit is the core element in the automation of the machining process. During the machining process, the control command and feedback signals are used to produce multiple real-time, in-process electronic data. This quantitative data gives a specific description of the working conditions and working status of the machine. The main source for sensing is the internal in-process data in the numerical controller. This data includes data received from interpolation, i.e. G-codes (e.g. speed feed, interpolation data, machining process, and cycle) and data from a servo device and motor (e.g. spindle power, feeding axes current, spindle current). The external data includes data received from sensors (e.g. vibrations, temperatures, visions) and process parameters (e.g. machining width and depth, and material removal rate). After receiving the internal and external data, sensing and connection take place by combining the internal and external data. A correlation between the machining process and feedback from the machine sets up independent sensing of the intelligent machine tool. Big data for the full life cycle of the machine

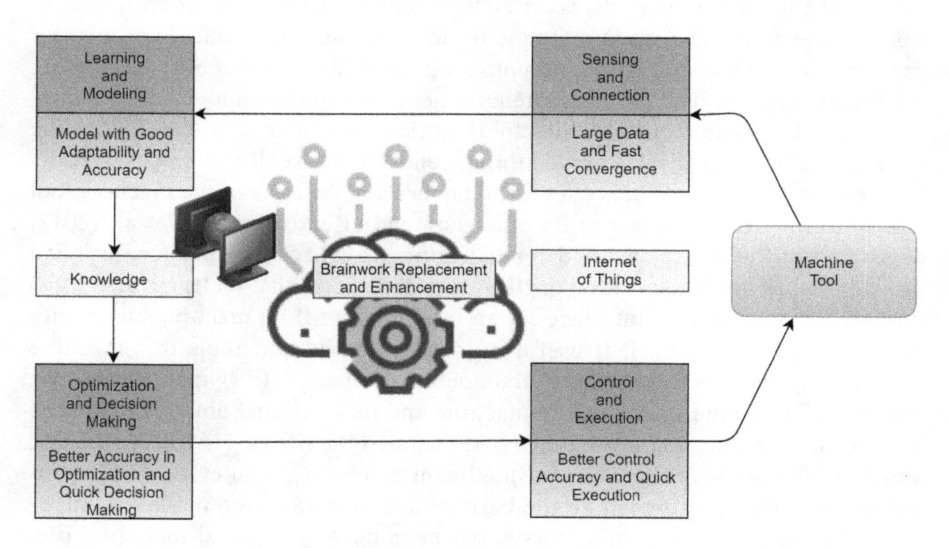

FIGURE 9.4 Intelligent machine tool system.

tool is established on the basis of the aggregation of big data from machining and the interconnections of the machine tool (Chen et al., 2019).

Independent learning and modelling in an intelligent machining tool generates knowledge through learning. The knowledge is generated through any one of three methods: theoretical modelling built on the physics of the machining process; big data modelling built on the correlation between the machining process and the status of the machine tool; and hybrid modelling built on theoretical analysis and big data from machining. The models can be used for similar machine tools to develop a digital twin of the machine tool (Chen et al., 2019).

During machining, the response of the machining operation can be predicted on the basis of previous outcomes. A model is developed on the basis of previous outcomes and is used to predict the response of the machining operation. Additionally, multi-objective optimisation is performed to improve the overall performance of the machining process in terms of product quality and productivity. An intelligent machine takes decisions based on intelligent codes specifically generated on the basis of optimum results. These codes are an important element in the optimisation and decision-making task. They give specific instructions to the instruction domain on the basis of multi-objective optimisation. Finally, the control and execution take place by integrating G-codes and intelligent codes. This results in a highly efficient, precise, reliable, safe, and low energy–consuming machining operation.

9.3 CUTTING TOOLS

A cutting tool is used in machining processes to remove unwanted material from a workpiece. Tool failure is a complicated phenomenon as it is governed by various parameters associated with machining. Intelligent machine tools can use smart cutting tools along with their real-time condition monitoring system. This may result in cost reduction, quality and productivity improvement, and damage prevention. Smart tooling for machining is one of the prominent technologies to achieve precision machining in the perspective of Industry 4.0. Smart cutting tools are composed of independent sensing and learning. These tools are capable of in-process sensing and actuation. Very few attempts have been made to design and develop smart cutting tools. Wang et al. (2014) designed a smart cutting tool for high value machining of hybrid material. The tool design is based on surface acoustic waves. Chen et al. (2014) developed a smart cutting tool for the turning operation in ultra-precision and micro cutting. The tool measures cutting forces and feed forces with the help of two pieces of piezoelectric film. Cheng et al. (2017) designed four smart cutting tools: force-based and temperature-based cutting tools, fast tool servo, and smart collets. Li et al. (2019) proposed a novel approach in which thin film thermocouples were embedded on the rake face of a tool. It was observed that the embedded thermocouples effectively monitor the temperature of the tool. Möhring et al. (2020) developed a smart milling tool as a cyber-physical system to improve the reliability and life of the tool. The monitoring of tool wear is performed by intelligent data evaluation.

9.3.1 Real-Time Tool Condition Monitoring

The complications in the machining process and the parameters associated with it make it difficult to predict the life of a cutting tool. Empirical models developed for predicting tool life, including Taylor's tool life equation, are not accurate due to the complexity associated with tool life. Effective real-time tool condition monitoring is monitoring of the tool state and prevention of workpiece damage. Worldwide researchers have applied their efforts in the domain of tool condition monitoring. Nouri et al. (2015) proposed a new method for monitoring an end mill cutter on the basis of force model coefficients. Bhuiyan et al. (2016) studied tool wear and plastic deformation through measuring the signal frequency in machining by using an acoustic emission sensor. Mali et al. (2017) proposed a system for monitoring a single point cutting tool for machining Ni/Cr/Mo alloy steel based on sensory data. Antic et al. (2018) proposed an approach for tool condition monitoring built on a texture-based description. Lee et al. (2019) proposed a method of intelligent tool condition monitoring. Sustainability-related manufacturing trade-offs and an optimum machining condition based on the machine tool state were identified by the system. Ochoa et al. (2019) designed a novel approach for tool monitoring on the basis of a stacked sparse autoencoder. Ou et al. (2021) developed a stacked denoising autoencoder neural network method for tool condition monitoring. The method is integrated with an online sequential learning machine. Denkena et al. (2020) developed an online feature assessment for tool condition monitoring based on wear curves. Gao et al. (2020) studied micro friction stir welding to monitor tool wear through a hybrid hierarchical spatiotemporal model. Bagga et al. (2020) implemented digital image processing techniques in a turning operation to measure two-dimensional flank wear measurement and monitoring of a tool. Xie et al. (2021) proposed a digital twin–driven data flow framework for monitoring the life cycle of a cutting tool. Hassan et al. (2021) suggested a sensor-related hybrid approach for tool condition monitoring in milling processes. Zhang et al. (2021) proposed a generic model to monitor wear in milling processes. He et al. (2021) used a cutting tool with a thin film thermocouple to predict tool wear in line with the deep learning method.

During machining, signals such as forces, vibrations, feed motor current, acoustic emissions, and spindle motor power are considered good indicators to detect tool failure. Figure 9.5 depicts general tool condition monitoring with adaptive control. The tool condition monitoring is done through trend analysis and/or pattern recognition. In trend analysis, the signal trends are analysed on the basis of the signal history. Signal patterns are determined for different tool conditions in pattern recognition. During tool condition monitoring, features extracted from the sensor feedback should be capable of identifying the tool condition under varying machining conditions including workpiece material and tool material. The extracted features should be distinguishable so that these features will be uniquely identified and not mixed up with irregularities in the process. In a tool condition monitoring system, monitoring and outlining tool failure play a crucial role in achieving sufficient flexibility to deal with an adaptive controlled process. The feedback signals received

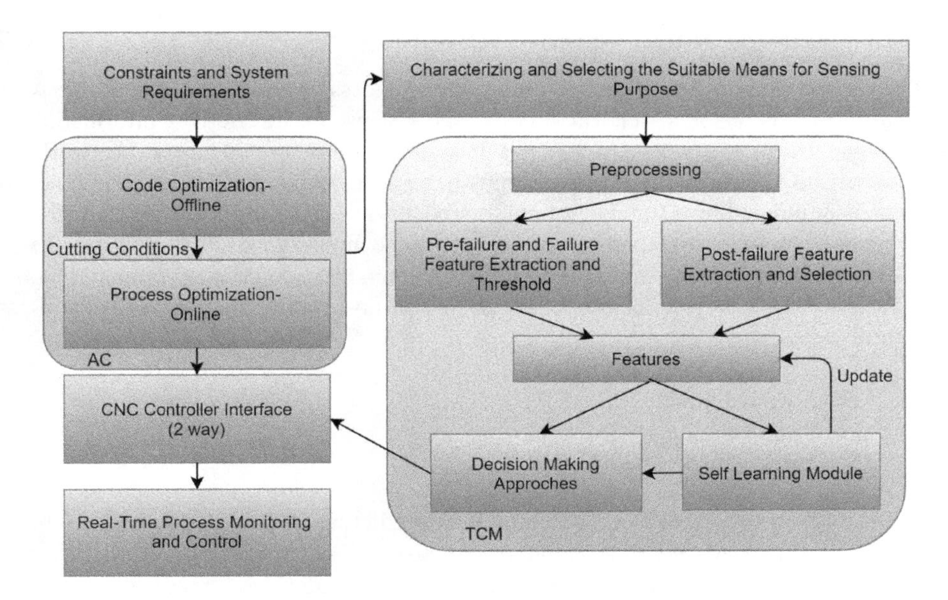

FIGURE 9.5 General tool condition monitoring system.

from multiple sensors will be processed by signal processing and a decision-making algorithm. The decision will be communicated to the machine controller to protect the workpiece. The output of the algorithms should not be influenced by machining parameters, workpiece parameters, or tool parameters (Hassan et al., 2018).

9.4 TECHNIQUES

The performance of a machining process depends on several parameters including machining, workpiece, and tool parameters. Therefore, analysing the performance of a machining process is a complicated task. In intelligent machining, sensors give feedback to the system. After receiving the feedback, the decision-making depends on the model of the machining process based on the physics of the process. The model of the machining process is typically in the form of a differential equation. However, due to the complex nature of the process the model becomes an inappropriate solution to the problem. In the case of sufficient data defining the process, a model based on the data can be developed. This model can be completed on the basis of soft computing techniques (Deb and Dixit, 2008). Soft computing is a computational paradigm that provides imprecise models. These models are further divided into two kinds of techniques, i.e. approximate reasoning and functional optimisation and random search. Soft computing techniques are tolerant of uncertainty, imprecision, and partial truths. The human brain/mind can be considered the role model for soft computing. The outcomes of soft computing models are more reliable for solving real-world problems as these techniques are adaptive.

9.4.1 Artificial Neural Network (ANN)

The human brain is composed of information processing units called neurons. A neuron consists of soma, dendrites, axons, and synapse for transferring information from one neuron to another. An ANN is an attempt to prototype the neural network. An artificial neuron receives input signals in terms of numerical values. The input value is multiplied by a suitable weight to model synapse action. The weighted sum of inputs is computed to represent the strength of the input signals. Further, a suitable activation function is applied to the sum of inputs in order to determine the output. Figure 9.6 shows a typical neuron in an ANN. The most commonly used ANN architectures are

a. Feedforward neural networks
b. Feedback neural networks
c. Self-organising neural networks

An ANN is useful in intelligent machining for predicting responses. Some researchers have used an ANN for developing predictive models in machining processes; for example, D'Addona et al. (2011) studied an ANN to develop a predictive model for tool wear in the turning of aircraft engine products made of Inconel 718. Zain et al. (2012) developed an ANN-based mathematical model for predicting surface roughness in the end milling of Ti–6Al4V. Singh and Misra (2019) applied an ANN in electrical discharge machining for predicting the recast layer and surface finish to machine Nimonic 263. Singh et al. (2020) applied an ANN in electrical discharge machining to develop a model for the material removal rate of aluminium 7075 alloy. Paturi et al. (2021) applied an ANN approach to estimate cutting tool wear in the machining of AISI 52100 steel. Sada and Ikpeseni (2021) studied an ANN and an adaptive neuro-fuzzy inference system for predicting the material removal rate and tool wear in turning AISI 1050 steel. Bagga et al. (2021) studied tool wear measurements based on an ANN in machining EN-8 using a cemented carbide insert. Trinath et al. (2021) studied the machinability of composite material manufactured by the stir casting method. A model based on an ANN has been developed to predict cutting forces under suitable machining parameters.

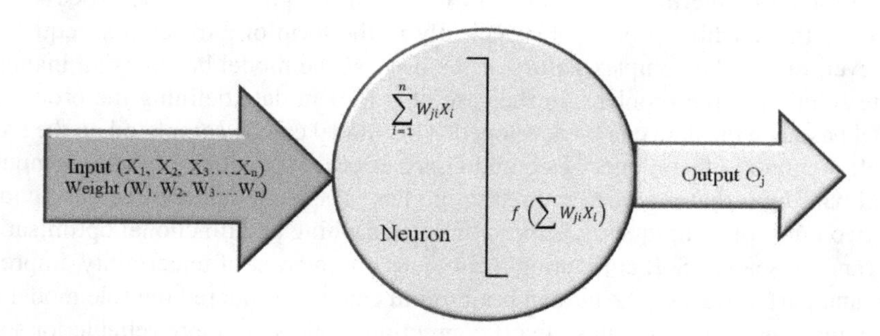

FIGURE 9.6 A typical neuron in an ANN.

In an ANN, neural networks are being trained to generate an appropriate response to the given inputs. Training is the process in which iterations are performed to adjust the weights and biases to the inputs. The iterations are performed until the network is able to generate the required output from the given inputs. The training process is categorised as supervised training and unsupervised training. The output is specified as Equation (9.1):

$$O_j = f\left(\Sigma W_{ji} X_i\right) \tag{9.1}$$

where:

O_j = output
W_{ji} = weight associated with jth neuron
X_i = input

9.4.2 Fuzzy Sets

Fuzzy logic is an essential component in soft computing. In 1965, Zadeh coined the term "fuzzy" for the logic working based on the vague set membership. Fuzzy logic is helpful to deal with uncertainty. The first commercial application using fuzzy logic was launched in 1981 for a control system. Fuzzy logic depends on fuzzy sets/fuzzy algebra. An interesting fact about fuzzy logic is that it essentially combines different algebras such as relational algebra, Boolean algebra, and predicate algebra. A fuzzy set is a set having degrees of membership between 1 and 0.

Definition of a fuzzy set:

If S is a universe of discourse and x Є S, fuzzy set X in S is defined as ordered pairs, i.e. $X = \{(x, g(x)) | x \in S\}$, where $g(x)$ is the membership value (as given in Equation 9.2).

- Example: X = set of all honest students.

$$X = \left\{\left(x, g\left(x\right)\right) \middle| x \in \mathrm{E}\right\} \tag{9.2}$$

where $g(x)$ is a measure of honesty.

X={(Reva, 0.7), (Kedar, 0.9), (Ranita, 0.5)}

Ramanujam et al. (2014) studied the turning of Inconel 625 to optimise the process parameters using fuzzy-based principal component analysis. Pathak et al. (2018) performed the optimisation of process parameters in the dry turning of AISI A2 tool steel using fuzzy logic. Nagaraju et al. (2018) studied material removal and surface finish in the machining of AISI 304 stainless steel. Further fuzzy logic was used for optimisation to improve the material removal rate and surface finish. Phate et al. (2019) studied the wire electrical discharge machining of Al/SiCp20 metal matrix composite material to optimise the process parameters using the grey-fuzzy approach. Upputuri et al. (2020) performed optimisation of the process parameters in drilling carbon fibre–reinforced polymer using fuzzy logic. Shirguppikar et al. (2020) used grey-fuzzy logic for optimising the electrical discharge machining of bright mild steel. Guha et al. (2020) studied the wire electric discharge machining of AISI4140 alloy steel to optimise the process parameters using regression analysis and the grey-fuzzy approach. Sanghvi et al. (2021) used the fuzzy logic–grey relation analysis approach to optimise the process parameters in the end milling of Inconel 825 with a coated tool. Pramanick et al. (2021) performed optimisation of the electrical discharge machining of spark plasma–sintered monolithic boron carbide using fuzzy logic and the multi-performance characteristics indices technique.

9.5 DEVELOPMENT TRENDS IN INTELLIGENT MACHINING

The ever-increasing demands for low cost, high quality products have forced the manufacturing industries to develop and adopt advanced practices in planning, control, and decision-making in traditional machining. The research in this domain has transformed traditional machining into intelligent machining. Intelligent machining is capable of manufacturing low cost and high quality products at faster production rates. In future, machine learning can be used for modelling the machining process based on the data accumulated for the development of intelligent machine tools in the context of Industry 4.0. In the domain of smart tooling and condition monitoring, smart jigs and fixtures can be developed. Also in tool condition monitoring, a reliable and robust general system can be developed to detect pre-failures of cutting tools. In the domain of soft computing techniques for the modelling and optimisation of machining processes, more robust and efficient computing methods and optimisation algorithms can be developed to meet the needs of Industry 4.0.

9.6 SUMMARY

A review and discussion on recent trends in intelligent machining are described in this chapter. The chapter is divided into three interrelated sections, i.e. machine tools, cutting tools, and techniques. The machine tools section is focused on intelligent machine tools used in intelligent machining. Cutting tools mainly refer to smart cutting tools and their real-time condition monitoring in intelligent machining. The techniques section is focused on soft computing techniques and optimisation techniques used in intelligent machining. Finally, the chapter summarises the development trends in intelligent machine tools, cutting tools, and optimisation techniques. Further research is needed in the domain of intelligent machining to develop systems based on the industrial internet of things to improve the efficiency, reliability, and sustainability of machining processes.

REFERENCES

Al-Maeeni, S. S. H., Kuhnhen, C., Engel, B., & Schiller, M. 2020. Smart retrofitting of machine tools in the context of industry 4.0. *Procedia CIRP*, 88:369–374.

Antić, A., Popović, B., Krstanović, L., Obradović, R., & Milošević, M. 2018. Novel texture-based descriptors for tool wear condition monitoring. *Mechanical Systems and Signal Processing*, 98:1–15.

Astakhov, V. P., & Outeiro, J. C. 2008. Metal cutting mechanics, finite element modelling. In *Machining*, ed. J. P. Davim, 1–27. Springer, London.

Bagga, P. J., Makhesana, M. A., Patel, H. D., & Patel, K. M. 2021. Indirect method of tool wear measurement and prediction using ANN network in machining process. *Materials Today: Proceedings*, 44:1549–1554

Bagga, P. J., Makhesana, M. A., Patel, K., & Patel, K. M. 2020. Tool wear monitoring in turning using image processing techniques. *Materials Today: Proceedings*, 44:771–775.

Barton, D., Gönnheimer, P., Schade, F., Ehrmann, C., Becker, J., & Fleischer, J. 2019. Modular smart controller for Industry 4.0 functions in machine tools. *Procedia CIRP*, 81:1331–1336.

Bhuiyan, M. S. H., Choudhury, I. A., Dahari, M., Nukman, Y., & Dawal, S. Z. 2016. Application of acoustic emission sensor to investigate the frequency of tool wear and plastic deformation in tool condition monitoring. *Measurement*, 92:208–217.

Botcha, B., Iquebal, A. S., & Bukkapatnam, S. T. 2020. Smart manufacturing multiplex. *Manufacturing Letters*, 25:102–106.

Chen, J., Hu, P., Zhou, H., Yang, J., Xie, J., Jiang, Y., & Zhang, C. 2019. Toward intelligent machine tool. *Engineering*, 5:679–690.

Chen, X., Cheng, K., & Wang, C. 2014. Design of a smart turning tool with application to in-process cutting force measurement in ultraprecision and micro cutting. *Manufacturing Letters*, 2:112–117.

Chen, Z. C., & Fu, Q. 2011. An optimal approach to multiple tool selection and their numerical control path generation for aggressive rough machining of pockets with free-form boundaries. *Computer-Aided Design*, 43:651–663.

Cheng, K., Niu, Z. C., Wang, R. C., Rakowski, R., & Bateman, R. 2017. Smart cutting tools and smart machining: Development approaches, and their implementation and application perspectives. *Chinese Journal of Mechanical Engineering*, 30:1162–1176.

Costa, S., Silva, F. J. G., Campilho, R. D. S. G., & Pereira, T. 2020. Guidelines for Machine tool sensing and smart manufacturing integration. *Procedia Manufacturing*, 51:251–257.

D'Addona, D., Segreto, T., Simeone, A., & Teti, R. 2011. ANN tool wear modelling in the machining of nickel superalloy industrial products. *CIRP Journal of Manufacturing Science and Technology*, 4:33–37.

Deb, Sankha, & Dixit, U. S. 2008. Intelligent machining: Computational methods and optimization. In *Machining*, ed. J. P. Davim, 329–358. Springer, London.

Denkena, B., Bergmann, B., & Stiehl, T. H. 2020. Wear curve based online feature assessment for tool condition monitoring. *Procedia CIRP*, 88:312–317.

Desforges, X., Habbadi, A., & Archimède, B. 2011. Design methodology for smart actuator services for machine tool and machining control and monitoring. *Robotics and Computer-Integrated Manufacturing*, 27:963–976.

Durmaz, Y. M., & Yildiz, F. 2019. The wear performance of carbide tools coated with TiAlSiN, AlCrN and TiAlN ceramic films in intelligent machining process. *Ceramics international*, 45:3839–3848.

Gao, Z., Chen, M., Guo, W. G., & Li, J. 2020. Tool wear characterization and monitoring with hierarchical spatio-temporal models for micro-friction stir welding. *Journal of Manufacturing Processes*, 56:1353–1365.

Guha, S., Das, P. P., & Routara, B. C. 2020. Parametric optimization of wire electric discharge machining on AISI4140 alloy steel using regression analysis and grey-fuzzy approach. *Materials Today: Proceedings*. https://doi.org/10.1016/j.matpr.2020.10.305

Gupta, E. V., Mogale, D. G., & Tiwari, M. K. 2019. Optimal control of production and maintenance operations in smart custom manufacturing systems with multiple machines. *IFAC-PapersOnLine*, 52:241–246.

Hassan, M., Sadek, A., & Attia, M. H. 2021. Novel sensor-based tool wear monitoring approach for seamless implementation in high speed milling applications. *CIRP Annals* 70 (1): 87–90. https://doi.org/10.1016/j.cirp.2021.03.024

Hassan, M., Sadek, A., Attia, M. H., & Thomson, V. 2018. Intelligent machining: real-time tool condition monitoring and intelligent adaptive control systems. *Journal of Machine Engineering*, 18:5–17.

He, Z., Shi, T., Xuan, J., & Li, T. 2021. Research on tool wear prediction based on temperature signals and deep learning. *Wear*, 478:203902.

Jankowski, M., & Wozniak, A. 2016. Mechanical model of errors of probes for numerical controlled machine tools. *Measurement*, 77:317–326.

Jiang, X., & Cheng, X. 2012. Method of humanity design for numerical control machine tool. *Energy Procedia*, 17:650–654.

Lee, W. J., Mendis, G. P., & Sutherland, J. W. 2019. Development of an intelligent tool condition monitoring system to identify manufacturing tradeoffs and optimal machining conditions. *Procedia Manufacturing*, 33:256–263.

Li, B., Zhang, H., Ye, P., & Wang, J. 2020. Trajectory smoothing method using reinforcement learning for computer numerical control machine tools. *Robotics and Computer-Integrated Manufacturing*, 61:101847. https://doi.org/10.1016/j.rcim.2019.101847.

Li, G., Zhou, H., Jing, X., Tian, G., & Li, L. 2017. An intelligent wheel position searching algorithm for cutting tool grooves with diverse machining precision requirements. *International Journal of Machine Tools and Manufacture*, 122:149–160.

Li, J., Tao, B., Huang, S., & Yin, Z. 2019. Cutting tools embedded with thin film thermocouples vertically to the rake face for temperature measurement. *Sensors and Actuators A: Physical*, 296:392–399.

Liu, C., Zhang, L., Niu, J., Yao, R., & Wu, C. 2020. Intelligent prognostics of machining tools based on adaptive variational mode decomposition and deep learning method with attention mechanism. *Neurocomputing*, 417:239–254.

Liu, M., Yao, Y., & Du, J. 2021. An efficient machine tool control instruction compression method for networked numerical control systems. *Robotics and Computer-Integrated Manufacturing*, 67:102027.

Liu, W., Kong, C., Niu, Q., Jiang, J., & Zhou, X. 2020. A method of NC machine tools intelligent monitoring system in smart factories. *Robotics and Computer-Integrated Manufacturing*, 61:101842.

Liu, Z., Chen, W., Zhang, C., Yang, C., & Cheng, Q. 2021. Intelligent scheduling of a feature-process-machine tool supernetwork based on digital twin workshop. *Journal of Manufacturing Systems*, 58:157–167.

Longo, F., & Padovano, A. 2020. Voice-enabled assistants of the Operator 4.0 in the social smart factory: Prospective role and challenges for an advanced human–machine interaction. *Manufacturing Letters*, 26:12–16.

Lv, J., Tang, R., & Jia, S. 2014. Therblig-based energy supply modeling of computer numerical control machine tools. *Journal of Cleaner Production*, 65:168–177.

Lv, J., Tang, R., Jia, S., & Liu, Y. 2016. Experimental study on energy consumption of computer numerical control machine tools. *Journal of Cleaner Production*, 112:3864–3874.

Mahboubkhah, M., Akhbari, S., & Barari, A. 2019. Self-configuration machining capability of a 4-DOF parallel kinematic machine tool with non-singular workspace for intelligent manufacturing systems. *IFAC-PapersOnLine*, 52:288–293.

Mali, R., Telsang, M. T., & Gupta, T. V. K. 2017. Real time tool wear condition monitoring in hard turning of Inconel 718 using sensor fusion system. *Materials Today: Proceedings*, 4:8605–8612.

Möhring, H. C., Werkle, K., & Maier, W. 2020. Process monitoring with a cyber-physical cutting tool. *Procedia CIRP*, 93:1466–1471.

Nagaraju, N., Venkatesu, S., & Ujwala, N. G. 2018. Optimization of process parameters of EDM process using fuzzy logic and Taguchi methods for improving material removal rate and surface finish. *Materials Today: Proceedings*, 5:7420–7428.

Nassehi, A., & Newman, S. T. 2012. Modeling of machine tools using smart interlocking . software blocks. *CIRP Annals*, 61:435–438.

Netzer, M., Michelberger, J., & Fleischer, J. 2020. Intelligent anomaly detection of machine tools based on mean shift clustering. *Procedia CIRP*, 93:1448–1453.

Nouri, M., Fussell, B. K., Ziniti, B. L., & Linder, E. 2015. Real-time tool wear monitoring in milling using a cutting condition independent method. *International Journal of Machine Tools and Manufacture*, 89:1–13.

Ochoa, L. E. E., Quinde, I. B. R., Sumba, J. P. C., Guevara Jr, A. V., & Morales-Menendez, R. 2019. New approach based on autoencoders to monitor the tool wear condition in HSM. *IFAC-PapersOnLine*, 52:206–211.

O'Driscoll, E., Kelly, K., & O'Donnell, G. E. 2015. Intelligent energy based status identification as a platform for improvement of machine tool efficiency and effectiveness. *Journal of Cleaner Production*, 105:184–195.

Ou, J., Li, H., Huang, G., & Yang, G. 2021. Intelligent analysis of tool wear state using stacked denoising autoencoder with online sequential-extreme learning machine. *Measurement*, 167:108153.

Paiva, P. R., de Freitas, B. I., Carvalho, L. K., & Basilio, J. C. 2021. Online fault diagnosis for smart machines embedded in Industry 4.0 manufacturing systems: A labeled Petri net-based approach. *IFAC Journal of Systems and Control*, 16:100146.

Pan, L., Guo, X., Luan, Y., & Wang, H. 2021. Design and realization of cutting simulation function of digital twin system of CNC machine tool. *Procedia Computer Science*, 183:261–266.

Park, H. S., Qi, B., Dang, D. V., & Park, D. Y. 2018. Development of smart machining system for optimizing feed rates to minimize machining time. *Journal of Computational Design and Engineering*, 5:299–304.

Pathak, A. D., Warghane, R. S., & Deokar, S. U. 2018. Optimization of cutting parameters in dry turning of AISI A2 tool steel using carbide tool by Taguchi based fuzzy logics. *Materials Today: Proceedings*, 5:5082–5090.

Paturi, U. M. R., Cheruku, S., Pasunuri, V. P. K., & Salike, S. 2021. Modeling of tool wear in machining of AISI 52100 steel using artificial neural networks. *Materials Today: Proceedings*, 38:2358–2365.

Pawanr, S., Garg, G. K., & Routroy, S. 2021. Modelling of variable energy consumption for CNC machine tools. *Procedia CIRP*, 98:247–251.

Phate, M. R., Toney, S. B., & Phate, V. R. 2019. Analysis of machining parameters in WEDM of Al/SiCp20 MMC using Taguchi-based grey-fuzzy approach. *Modelling and Simulation in Engineering*, 2019:1483169. https://doi.org/10.1155/2019/1483169

Pramanick, A., Mandal, S., Dey, P. P., & Das, P. K. 2021. WEDM process optimization of sintered structural ceramic sample by using fuzzy-MPCI technique. *Materials Today: Proceedings*, 41:925–934.

Ramanujam, R., Venkatesan, K., Saxena, V., Pandey, R., Harsha, T., & Kumar, G. 2014. Optimization of machining parameters using fuzzy based principal component analysis during dry turning operation of Inconel 625: A hybrid approach. *Procedia Engineering*, 97:668–676.

Ramesh, R., Jyothirmai, S., & Lavanya, K. 2013. Intelligent automation of design and manufacturing in machine tools using an open architecture motion controller. *Journal of Manufacturing Systems*, 32:248–259.

Sada, S. O., & Ikpeseni, S. C. 2021. Evaluation of ANN and ANFIS modeling ability in the prediction of AISI 1050 steel machining performance. *Heliyon*, 7(2):e06136.

Sanghvi, N., Vora, D., Patel, J., & Malik, A. 2021. Optimization of end milling of Inconel 825 with coated tool: A mathematical comparison between GRA, TOPSIS and fuzzy logic methods. *Materials Today: Proceedings*, 38:2301–2309.

Schmid, J., Schmid, A., Pichler, R., & Haas, F. 2020. Validation of machining operations by a virtual numerical controller kernel based simulation. *Procedia CIRP*, 93:1478–1483.

Schmucker, B., Trautwein, F., Semm, T., Lechler, A., Zaeh, M. F., & Verl, A. 2021. Implementation of an intelligent system architecture for process monitoring of machine tools. *Procedia CIRP*, 96:342–346.

Shen, W., Hu, T., Yin, Y., He, J., Tao, F., & Nee, A. Y. C. 2020. Digital twin based virtual commissioning for computerized numerical control machine tools. In *Digital Twin Driven Smart Design*, eds. F. Tao, A. Liu, T. Hu, & A. Y. C. Nee, 289–307. Academic Press, Cambridge, Massachusetts, United States.

Shirguppikar, S. S., Patil, M. S., & Ghorapade, V. 2020. Grey fuzzy multiobjective optimization of process parameters for dry electro discharge machining process. *Materials Today: Proceedings*, 27:671–676.

Singh, A. K., Singhal, D., & Kumar, R. 2020. Machining of aluminum 7075 alloy using EDM process: An ANN validation. *Materials Today: Proceedings*, 26:2839–2844.

Singh, B., & Misra, J. P. 2019. Surface finish analysis of wire electric discharge machined specimens by RSM and ANN modeling. *Measurement*, 137:225–237.

Trinath, K., Aepuru, R., Biswas, A., Viswanathan, M. R., & Manu, R. 2021. Study of self lubrication property of Al/SiC/Graphite hybrid composite during machining by using artificial neural networks (ANN). *Materials Today: Proceedings*, 44: 3881–3887.

Upputuri, H. B., Nimmagadda, V. S., & Duraisamy, E. 2020. Optimization of drilling parameters on carbon fiber reinforced polymer composites using fuzzy logic. *Materials Today: Proceedings*, 23:528–535.

Vieler, H., Lechler, A., & Riedel, O. 2017. Architecture and implementation of an interface for intelligent tools in machine tools. *Procedia Manufacturing*, 11:2077–2082.

Wang, C., Cheng, K., Minton, T., & Rakowski, R. 2014. Development of a novel surface acoustic wave (SAW) based smart cutting tool in machining hybrid dissimilar material. *Manufacturing Letters*, 2:21–25.

Wang, J., & Guo, J. 2019. The identification method of the relative position relationship between the rotary and linear axis of multi-axis numerical control machine tool by laser tracker. *Measurement*, 132:369–376.

Wei, Y., Hu, T., Zhang, W., Tao, F., & Nee, A. Y. C. 2020. Digital twin driven lean design for computerized numerical control machine tools. In *Digital Twin Driven Smart Design*, ed. F. Tao, A. Liu, T. Hu, & A. Y. C. Nee, 289–307. Academic Press, Cambridge, Massachusetts, United States.

Xie, Y., Lian, K., Liu, Q., Zhang, C., & Liu, H. 2021. Digital twin for cutting tool: Modeling, application and service strategy. *Journal of Manufacturing Systems*, 58:305–312.

Zain, A. M., Haron, H., Qasem, S. N., & Sharif, S. 2012. Regression and ANN models for estimating minimum value of machining performance. *Applied Mathematical Modelling*, 36:1477–1492.

Zhang, Y., Zhu, K., Duan, X., & Li, S. 2021. Tool wear estimation and life prognostics in milling: Model extension and generalization. *Mechanical Systems and Signal Processing*, 155:107617.

Zhao, F., Zhang, C., Yang, G., & Chen, C. 2016. Online machining error estimation method of numerical control gear grinding machine tool based on data analysis of internal sensors. *Mechanical Systems and Signal Processing*, 81:515–526.

Zhou, J., Li, P., Zhou, Y., Wang, B., Zang, J., & Meng, L. 2018. Toward new-generation intelligent manufacturing. *Engineering*, 4:11–20.

10 Digital Market Scenario in India

A Case Study on "Unicorn" Indian Digital Start-Ups

Sayak Pal and Nitesh Tripathi

CONTENTS

10.1 INTRODUCTION: UNDERSTANDING DIGITAL ECONOMY

The digital economy is changing the characteristics of India's information and communications technology and is also fostering economic and social change. Several factors such as research and development, laws, policies, and product and service development are also linked with this change in the economy (Brynjolfsson and Kahin 2002). In some of the available literature, the term digital economy was interpreted as a dynamic efficiency consisting of new products and activities. The digital economy is also referred to as the internet economy, the new economy, and the web economy, with the involvement of new technology resulting in a combination of new ideas. The new economy distinguishes between information and knowledge in order to segregate a set of data from structures that enables an organisation to interpret the data. While the old economy relied on physical forms of information exchange such as cash and check transactions, invoices, bills, reports, maps, photographs, direct mail, and similar technologies, the digital or new economy stores trillions of bytes of information in databases and shares them at lightning speed across nations (Carlsson 2004). The facilitation provided by the digital economy has three components: "supporting infrastructure", "electronic business processes", and "electronic commercial

DOI: 10.1201/9781003200857-10

transactions". The common factor that is fundamental to all three components is the "computer-mediated network" that makes the processes smoother and faster. The infrastructural support it provides includes sectors such as software, hardware, telecommunications, support services, and the human capital required for electronic business. Businesses that are facilitated include online product and services procurement, production management, logistics and supply chain management, vendor management, inventory management, and electronic auctions. The third most important component offers monetary transactions for various kinds of goods and services sold over an electronic medium. Measurement of the facilities offered by the digital economy is also easier because of the nature of the transactions that leave their footprint on databases. Multiple strategies are available for measurement such as "achieve and exploit first mover status", "leverage our core competencies", and "contract for e-business process expertise". The leading economies of the world release report on the contribution and growth of their digital economy. For example, the United States released "E-Stats E-Commerce 1999" which includes industries such as manufacturing, retail and wholesale trade as well as selected service industries. Much research has also been carried out on the digital economy and its application, for example Adam Fein's study on "changing supply chain industries and organisations" or Hal Varian's assessment of "how well NAICS captures e-business activities" (Mesenbourg 2001).

10.2 OBJECTIVES AND METHODS

The objective of the study is to assess various initiatives launched by the Government of India that aided the growth of digital start-ups in India. The study also includes various start-ups that became unicorns under the government's initiatives and support.

For this purpose, the case study method was utilised to collect information on the government's initiatives and strategies that helped foster the digital economy in India. The researcher conducted a secondary data analysis by consulting the literature relevant to the topic. Using articles, research papers, and research reports related to the digital economy, unicorns of India were studied in detail to make meaningful conclusions regarding the digital start-up scenario in India.

10.3 DIGITAL ECONOMY OF INDIA

The digitisation of the Indian economy aims to bring change to the lives of people living in India, which was an agenda of the Government of India to empower the nation digitally with three major outcomes: "faceless", "paperless", and "cashless". Another aim of this initiative was to popularise digital learning among people to enlist the nation into the informed economies of the world. The most daring initiative taken by the Indian government was "demonetisation" that created a ripple effect among people while rejecting the old currency and accepting new currencies (Sen 2020). India has made efforts to digitise its economy by limiting the use of physical cash transactions with the aim to eliminate black money from the economy. Before digitisation, India mostly used physical cash for transactions. According to Uma Shankar, the ratio of cash to the Indian gross domestic product (GDP) was 12.42% in 2004, which was considered one of the highest usages of cash in an economy and, for the same reason, the currency distribution rate of the country was also higher than most other nations. In a comparison between India and the United States in 2012–2013, the distribution of currency notes showed 76.47 billion in India while the United States had only 34.5 billion in circulation in their economy, showing the massive use of physical currency in the nation's economy before digitisation (Shankar 2017).

Demonetisation forced Indians to use a digital medium for various kinds of transactions and encouraged people to adapt to digital payments for professional tax, property tax, electricity and gas bills, online booking, registration and renewal of birth and death certificates, membership registration and renewals, fee payments, and much more. The shortage of physical currency, on the other hand, popularised digital cashless services such as internet banking, mobile banking, m-wallet, unified payment interface (UPI), Aadhaar enabled payment system (AEPS), e-books, e-commerce, e-shopping, and digital library. While the youth of India were the most tech savvy category of the economy, digitisation pushed other categories of people to adapt to the changes and make their transactions online. While the initiative was intended to boost the economy with a controlled and traceable system, the implications of this strategy are still questionable due to its effect on every Indian individual. A research study by Dishani Sen based on a semi-structured interview with 80 participants in the 25–55 years age group from rural and urban populations of West Bengal describes four key themes – "Aadhaar enabled payment system", "influence", "accessibility", and "social media" – which helped in understanding the impact of the "digital economy" on the Indian population. The result of the research presented a variety of responses showing that there were both positive and negative perceptions towards digitisation as well as growing concerns that deterred people from its implementation (Sen 2020).

In his study, Uma Shankar described the various benefits of the digitised economy, including reduced cases of tax avoidance; the elimination of black money; controlled price inflation for the real estate market; limited currency issuances; reduced usage of the physical infrastructure for the banking sector; more support for welfare programmes; limited circulation of fake currencies while maintaining hygiene during transactions by avoiding soiled and damaged currencies; reduced number of ATMs aimed at reducing the liabilities of banks; speedy transactions; and increased GDP. The study also described the challenges of the digital economy ranging from currency denomination; too much use of physical currencies; heavy use of ATM cards for currency withdrawal; limited availability and poor transaction culture of point of sale (POS) terminals; weaker mobile operations and connections (especially among rural areas); unavailability of resources for small businesses; poor perception of users towards digital transactions; fear of fraud; and lack of efficient payment gateways. Although many initiatives are in favour of making digitisation successful, the Indian economy has a long way to go before it can claim to be called an informed economy (Shankar 2017).

10.4 MAKE IN INDIA INITIATIVE (2014)

In 2014, the Make in India initiative was launched by the prime minister to strengthen the economic growth of the nation by strengthening the manufacturing sectors across the country. The primary aim of the Make in India initiative was the revival of the manufacturing sectors for long-term economic development and protection of basic production units such as minerals, power, and water with a competitive approach. The Make in India initiative not only included industries ranging from agriculture to mining and manufacturing but also considered different services. A total of 25 economic sectors were identified under the initiative: automobiles; aviation; biotechnology; construction; defence manufacturing; electrical machinery; food processing; automobile components; ports and shipping; information technology and business process management; media and entertainment; leather; chemicals; electronic systems; oil and gas; pharmaceuticals; railways; renewable energy; mining; roads and highways; space and astronomy; textiles and garments; thermal power; tourism and hospitality; and wellness (Nam and Steinhoff 2018).

According to data from Statista, the annual production growth rate of the manufacturing industry in India from 2013 to 2019 was steady and coincided with the implementation of the Make in India initiative (3.6% in 2014; 3.8% in 2015; 2.8% in 2016; 4.4% in 2017; 4.8% in 2018; and 3.9% in 2019) (Statista 2019). On the other hand, the Ease of Doing Business (EoDB) Index suggests better and simpler regulations for business along with stronger property rights. The Make in India initiative covers 12 areas of regulation related to manufacturing businesses, ranging from "protecting minority investors"; "starting a business of all"; "paying taxes"; "dealing with construction permits"; "enforcing contracts"; "resolving insolvency"; "contracting with the government"; "trading across borders"; "getting credit"; "registering property"; "getting electricity"; and "employing workers" (Government of India 2014). According to the data published on India's performance on Ease of Doing Business, the nation has significantly improved its ranking in this global parameter from 132 in 2010 to 63 in 2019, showing the opportunities that lie ahead for businesses to grow (The World Bank 2019).

Efforts made through the Make in India initiative also aimed to improve the nation's World Bank ranking from 142 to 50 among 149 countries while seeking larger amounts of foreign investment from other nations. Make in India includes 25 different sectors, the focus was to develop the most prominent industries such as aviation, textile, information technology, automobile, construction, etc. The plan also included the conversion of five smart cities to generate more industry clusters leading to an increased GDP and more jobs. The Make in India initiative was launched to boost India's manufacturing capacity, making the nation host to world-class infrastructure while asking manufacturing businesses across the world to set up their bases in India, which would also create job opportunities for citizens. According to observations made by Perwez, Vijayalakshmi, and Perwaiz (2016), the initiative focuses on a strong determination, a clear vision, vast resources, efficient administration, and detailed planning, which would help in employing more people by creating opportunities. However, certain forces are preventing the government from achieving this monumental task, which they identified as five focus areas: "incentivising utilisation"; "defence production and localisation"; "capacity utilisation"; "power sector"; and "encouraging alternative technologies" (Perwez, Vijayalakshmi and Perwaiz 2016).

The Make in India initiative has faced many challenges with the 10 most crucial challenges being "improving the level of education and training"; "supporting entrepreneurial growth by reforming labour regulations"; "efficient land acquisition"; "reforming taxation related to manufacturing"; "improving and enhancing the speed of regulatory approvals"; "working towards user-friendly bilateral investment treaties (BITs)"; "removing high tariff and non-tariff barriers while at the same time providing trade facilitation"; "eliminating infrastructure bottlenecks"; "dealing with non-performing assets (NPAs) as regards public sector banks", and "removing endemic corruption at all levels" (Som 2018).

10.5 DIGITAL INDIA CAMPAIGN (2015)

The Digital India campaign was a major initiative of the Government of India. Launched on 1 July 2015, the aim of the campaign is to shift the Indian economy towards an informed economy backed by good governance offering participation, transparency, and responsiveness to its citizens. The initiative has three major goals that are also considered the visions of the programme: "creating infrastructure for every Indians" refers to the availability of high speed internet connection for every citizen, encouraging them to participate in digital and financial transactions; "digital empowerment of citizens" aims to empower citizens and provide them with digital literacy and access to digital resources, leading to cashless transactions; and "on demand governance and services" looks forward to providing flawless and prompt services to people through online and mobile channels, maintaining harmonious integration across government departments and jurisdictions. The Digital India campaign undertook several projects with the aim of improving the economic supremacy of the nation, such as the Digital Locker system to limit the use of paper; the launch of MyGov.in for e-governance; the Swachh Bharat mobile application; the e-signing facility using Aadhaar authentication; the e-hospital facilitation for online registration, booking, fee payment, booking a doctor's appointment, checking availability, online diagnostics, etc.; the National Scholarship Portal for e-submissions, sanctions, and disbursements; the Bharat Net programme to connect 250,000 village Panchayats, which was the largest rural broadband project undertaken by the Government of India; BSNL's Next Generation Network to provide the new generation communication services; the Outsourcing Policy for the benefit of north-eastern as well as smaller towns of India; the Electronics Development Fund to encourage and nurture innovation, research, and development throughout the country to establish a self-sustaining environment; the funding of the National Centre for Flexible Electronics to encourage research and innovation in emerging areas; empowering post-offices as multi-service support centres; connecting schools with internet facilities; and many more (Mohanta, Debasish and Nanda 2017).

The Digital India initiative aims to empower the nine pillars of economic growth for the nation starting with the "broadband highway" which aims to provide broadband internet services for all rural and urban areas, including 250,000 village Panchayats under the National Optical Fibre Network (NOFN); Universal Access to Mobile Connectivity involving the Department of Telecommunications, which would provide 55,619 villages with mobile connection to strengthen the communication processes of the nation; the Public Internet Access Programme which includes post offices as multi-service support centres to offer e-services with the help of the CSC Portal, DeitY's weblink, and the e-book portals; e-governance under the Digital India initiative aims to simplify government processes across departments using information technology to make government services more efficient; e-Kranti, which received approval from the Union Cabinet in 2015, came with the vision of "transforming e-governance for transforming governance" to promote mobile and good governance across the nation; and Information for All would allow people across India to use and reuse data from ministries and departments. Under the scheme, government will increase the interaction with people through web-based and social media platforms while a dedicated portal MyGov.in was launched in 2014 to help in better e-governance; "electronic manufacturing" focuses on encouraging and promoting electronic manufacturing within the nation to reduce imports to "zero" with the aim of providing a self-sufficient ecosystem; "IT for jobs" looks forward to training citizens in demanding skills to acquire the available jobs in the booming IT sector, which will provide more economic stability for the nation; the ninth pillar, the Early Harvest Programme, includes projects that need to be implemented within a short period of time, such as "Wi-Fi in all universities", "secure email within government", "school books to be eBooks", "public Wi-Fi hotspots" among others (Government of India 2015).

10.6 START-UP INDIA (2015)

Start-up India includes relatively new business initiatives in innovation, deployment, commercialisation, and development that have been running for no more than 5 years with less than 250 million rupees of turnover. The Government of India declared 2010–2020 the "decade of innovation" to foster the growth of science, technology, and innovation and also encourage entrepreneurship throughout the country. In 2016, the prime minister of India announced an action plan under the Start-up India initiative to encourage start-ups in India with various schemes ranging from Fund of Funds support with 100,000 million rupees, one-day company registration, relaxation of taxation on profits for the first three years, a tax exemption from capital gains tax, a guaranteed credit scheme, a faster and easier exit policy, zero capital gains on money that is invested in another start-up, self-certification for labour and environmental–related laws, the establishment of a Start-up India hub for relevant clearances, launching mobile apps, launching a portal for entrepreneurs to register, establishing friendly relationships between entrepreneurs and government, a reduction in patent free up to 80% and increasing the speed of patent examinations by implementing new intellectual property rights, government purchase for start-ups, facilitating women entrepreneurs with special schemes, providing assistance for start-ups in biotechnology, encouraging school students to partake in innovation, setting up "innovation centres" at national institutes, and building up a "research park" (Sarkar 2016).

Start-up India was launched on 15 August 2015 by the prime minister of India with the slogan "Start-up India, Stand up India" to encourage innovators to create opportunities in multiple areas with lucrative financial support and incentives (Jain 2016). A Start-up India hub was created with the three main objectives of setting up a beneficial collaboration between state and central government, investors from India and abroad, angel networks as well as between the financial institutions; focusing on aspects such as structuring businesses, marketing skills, feasibility testing, etc.; and arranging mentorship programmes involving government organisations. The initiative also included several silent features such as providing loans to help people set up businesses, setting up a network consisting of start-ups around the nation, encouraging youth entrepreneurship, selecting employees for low skilled jobs based on an interview, bringing transparency to the recruitment process, and providing special incentive packages to increase employment generation. A few programs were also identified to help boost start-ups in India such as the e-Biz portal introduced in 2015 to encourage young entrepreneurs with faster clearance; the Mudra Bank to provide financial support to people with micro-financial requirements; and the Atal Innovation mission to help innovators across academics, entrepreneurs, and researchers. Considering the huge number of young ideas struggling to establish in the market, Start-up India assists through multiple schemes, empowering youth entrepreneurs throughout the nation (Sarkar 2016).

10.7 ATMANIRBHAR BHARAT (2020)

Atmanirbhar Bharat, rooted in "self-resilient India", was announced by the prime minister of India on 12 May 2020; it provides economic support of 20 lakh crore with the aim to convert India into a self-resilient nation. To support the initiative of Atmanirbhar Bharat, all four major elements – labour, land, law, and liquidity – were essentially considered while covering home and cottage industries, micro, small, and medium enterprises (MSMEs), farmers, labourers, and all those industries that are taking action to boost the Indian economy. The Atmanirbhar Bharat initiative works to develop its five essential pillars starting with the "economy" which looks forward to taking a quantum leap instead of incremental changes; "infrastructure" aims to advance in order to construct an identity for the nation; "system", to fulfil the milestones of 21st-century India; "demography" to be used as an advantage for providing the energy to transform into a self-resilient nation; and "demand", to be harnessed in order to maintain and grow the demand and supply cycle. The initiative is divided into five phases: Phase I includes MSMEs; Phase II looks after the poor and migrant farmers; Phase III considers the agriculture sector; Phase IV concentrates on the growth of new horizons; and Phase V focuses on government reforms (Kamat and Kamat 2020).

The announcement of the Atmanirbhar Bharat initiative also created some misunderstandings, the most common of which was the complete cessation of imports into the nation; however, Atmanirbhar Bharat actually stressed reducing imports into

India while expanding exports to strengthen the nation's economy, which was also one of the prime objectives of the Make in India initiative launched in 2014, which focused on the growth of India's manufacturing units. Among its crucial packages, Atmanirbhar Bharat extends supports to the development of businesses, including 3 billion rupees of loans for the working capital of MSMEs and other businesses; 2 billion rupees of additional Kisan Credit Card support for farmers; affordable home loans for mid-income groups; 1.5 billion rupees of infrastructural support for agriculture and allied sectors; 0.08 billion rupees to "viability gap funding"; and 0.4 billion rupees support towards the 100 days employment scheme aimed at rural citizens under the Mahatma Gandhi National Rural Employment Guarantee Scheme (MGNREGS). The funding and other incentives under Atmanirbhar Bharat also aim at lifting India's Ease of Doing Business ranking to 50, which was also considered in the Make in India initiative (Massand, Lodi and Ambreen 2020).

Figure 10.1 shows the model proposed by Kamat and Kamat and the relationship between the 4Es (Education, Employability, Employment, and Entrepreneurship) and the five pillars of the Atmanirbhar Bharat initiative. As education is the instrument of development and transformation of a society, the Government of India struggled to decrease the number of school dropouts through various schemes, and implemented PM e-VIDYA which aims to collaborate online or on-air or through digital education with various multi-mode access portals such as DIKSHA (one nation-one digital platform); SWAYAM (online MOOC courses); IITPAL (for IITJEE/NEET preparation); CBSE Shiksha Vani podcast (an educational community radio service); and NIOS (educational support on sign language). The Government of India's employability-enhancing initiatives included several training programmes for the young generation

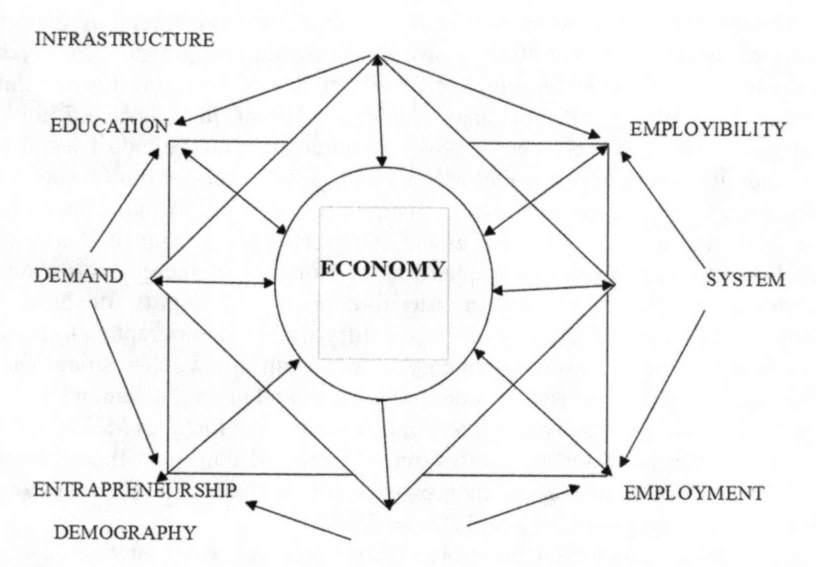

FIGURE 10.1 Proposed model of 4Es and five pillars of Atmanirbhar Bharat (Kamat and Kamat 2020).

to develop skills, such as the Skills Development Initiative (SDI) scheme for MSMEs; Skill Acquisition and Knowledge Awareness for Livelihood (SANKALP) for the disadvantaged section of society aligned to the objectives of the National Skill Development Mission (NSDM); Skills Strengthening for Industrial Value Enhancement (STRIVE) providing relevant and efficient skills development programmes through industrial training institutes (ITIs); Pradhan Mantri Kaushal Vikas Yojana (PMKVY) to train Indian youth for better employment; and Deen Dayal Upadhyaya Grameen Kaushalya Yojana (DDU-GKY) aiming to provide jobs to rural youth through skills development. While focusing on different skills development programmes, the Government of India is constantly working to provide employment to skilled individuals; however, India has a long way to go to eradicate unemployment. Encouraging entrepreneurship has already been a priority for the government through previous initiatives such as Start-up India, which aims to facilitate start-ups to foster the growth of the sector; however, the nation has yet to achieve a significant score on "self-resilience" by implementing multidimensional initiatives (Kamat and Kamat 2020).

10.8 UNICORNS

Aileen Lee, founder of the US-based venture capital firm Cowboy Venture, coined the term "unicorn" in 2013, in an article titled "Welcome to The Unicorn Club: Learning from Billion-Dollar Start-Ups". The term is given to companies based on their growth opportunities as well as their expected development. Three common categories exist under which companies are listed based on their valuation: unicorn includes businesses valued at US$1 billion and over; decacorn includes businesses valued at US$10 billion and over, while hectocorn includes businesses valued at US$100 billion and over (Alpha JWC Ventures 2020).

According to the "India Venture Capital Report 2021" by Bain & Company, India was counted among the top five countries in the world for its growth of 7000 start-ups in 2020 (Sheth, Krishnan and Upmanyu 2021). Table 10.1 provides details of the total number of unicorn start-ups (2020) in various sectors such as entertainment, e-commerce, hospitality, logistics, education, and food.

The highest valued Indian unicorns are brands such as One97 Communications ($16 billion), BYJU's ($13 billion), OYO ($9 billion), Ola Cabs ($6.3 billion), Zomato ($5.4 billion), and Swiggy ($5 billion), while a few brands also made it to the list in 2021, such as CRED ($2.2 billion), Messho ($2.1 billion), and ShareChat ($2.1 billion). The top contributors are One97 Communications who created Paytm Mall followed by BYJU's and OYO in second and third position, respectively (Buchholz 2021), and 2021 might witness some more soonicorns (having a business valuation below US$1 billion), such as Practo, Eruditus, Grofers, BharatPe, Vedantu, Ninjacart, Mswipe, MobiKwik, Capital, Navi, MPL, Licious, FreshToHome, Lendingkart, BankBazaar, Open, 1mg, Shadowfax, cure.fit, and Ecom Express, whose business values range from US$200 to 904 million, to enter into the unicorn camp of India (Chengappa 2021).

A high number of Indian soonicorns are showing massive potential in their business valuations to transform into unicorns, operating in the Indian marketplace.

TABLE 10.1
Unicorns of India

Consumer Tech		SaaS	
InMobi	Advertisements	ThoughtSpot	Horizontal business software
BYJU's	Edtech	Freshworks	Horizontal business software
Unacademy	Edtech	Icertis	Horizontal business software
Swiggy	Foodtech	HighRadius	Horizontal business software
Zomato	Foodtech	Druva	Horizontal infra software
Dream11	Gaming	Postman	Horizontal infra software
Snapdeal	Horizontal e-commerce	Zenoti	Vertical specific business software
Paytm Mall	Horizontal e-commerce	**Fintech**	
Dailyhunt	Media and entertainment	PolicyBazaar	Insurtech
Glance	Media and entertainment	Paytm	Payments
Ola	Mobility	Pine Labs	Payments
Ola Electric	Mobility	BillDesk	Payments
Oyo	Hospitality	Razorpay	Payments
BigBasket	Verticalised e-commerce	Zerodha	Wealth management
FirstCry	Verticalised e-commerce	**Others**	
Lenskart	Verticalised e-commerce	Greenko Group	Energy
Cars24	Verticalised e-commerce	ReNew Power	Energy
Nykaa	Verticalised e-commerce	Udaan	B2B marketplace
Shipping			
Delhivery	Shipping and logistics		
BlackBuck	Shipping and logistics		
Rivigo	Shipping and logistics		

Source: Data from Sheth, Krishnan, and Upmanyu (2021).

Topping the list is Practo, which facilities people with medical consultation valued at around US$904 million followed by BharatPe, a QR-based all payment app with an estimated value of US$885 million; Eruditus, an executive education provider valued at around US$850 million; Rebel Foods, the largest "cloud kitchen restaurant" chain in India with an estimated value of around US$799 million; and cure.fit, a health and fitness company providing online and offline services on multiple health and fitness-related issues, is also close to becoming a unicorn with an approximately US$775 million valuation. The chances of these businesses becoming unicorns are high with the help of government interventions towards transforming the Indian economy into a digitised economy (Usmani 2021).

10.9 CONCLUSION AND DISCUSSION

A study was carried out on the digital start-ups scenario in India where attempts were made to understand how the digital economy fostered the growth of these start-ups and ultimately turned them into companies worth billions of dollars. The power of digital technologies allows the rapid transfer of information that leads to faster transactions which ultimately means more business and revenue in a shorter time frame for organisations. It acts as a facilitator in sharing large volumes of information related to commerce leading to efficient business practices. This chapter began by delineating how government schemes/initiatives helped in boosting the digital infrastructure of the country and aided the incubation of digital start-ups that became market leaders. Most prominent initiatives such as Digital India, Make in India, Start-up India, Atmnirbhar Bharat, and a few others created a suitable environment for digital start-ups to flourish. The government gave a digital push to their functioning and also insisted that its citizens go digital in order to take advantage of government schemes. This, in turn, made people more receptive to the idea of becoming more "digital".

Most developed countries such as the United States have big multinational companies (MNC) that control the majority of the country's market (and also the world market). But in India, micro-, small- and medium-scale businesses (including the informal sector, cooperative societies, and cottage industries) have always been the biggest contributors to the nation's economy. Most sectors of the Indian market are not monopolised and thus provide ambient opportunities for micro-, small-, and medium-scale start-ups to increase the volume of their business and grow to their full potential. This is also true for the digital market in India as the government and the environment in general are encouraging for home-grown start-ups to expand and make their business go global. All of this has been possible due to the government's generous and spontaneous support and the democratisation of the market.

In recent years, the government has placed much emphasis on improving India's digital infrastructure as it is one of the best ways to boost the country's economy. Be it launching the Digital India campaign (2015), the Make in India initiative (2014), Start-up India (2015), Atmanirbhar Bharat (2020), and many more similar initiatives, there has been a huge surge in government efforts towards growing the digital economy. This is the reason why many digital innovations/start-ups have grown into big organisations and are providing employment for thousands of people. Several start-ups have even become big enough to take on much bigger foreign competitors be they at home or abroad. Start-ups such as Flipkart, Snapdeal, Zomato, Swiggy, Paytm, Ola, Unacademy, and Byju's have created ripples in the digital market and have paved the way for other Indian start-ups to flourish. A few of them, such as Ola, Byju's, and Unacademy, started from humble beginnings and eventually became unicorns (some even decacorns) in their sectors. To sum up, it can be said with certainty that government schemes and initiatives have been successful in boosting the digital infrastructure of India and aided the incubation of digital start-ups that are becoming market leaders and strong competitors for foreign companies operating in the Indian market. Small-scale digital start-ups have become huge brands with massive market shares in their respective domains within a short time span due to these very initiatives.

REFERENCES

Alpha JWC Ventures. 2020. The Differences Between Unicorn, Decacorn, and Hectocorn. Kuningan. www.alphajwc.com/en/the-differences-between-unicorn-decacorn-and-hectocorn/ (accessed 10 May 2021).

Brynjolfsson, Erik and Brian Kahin. 2002. "Introduction". In *Understanding the Digital Economy: Data, Tools, and Research*, by Brian Kahin and Erik Brynjolfsson, 12. Cambridge: The MIT Press.

Buchholz, Katharina. 2021. India's Unicorns. April 14. www.statista.com/chart/18183/highest-funded-startups-in-india/ (accessed 10 May 2021).

Carlsson, Bo. 2004 "The Digital Economy: What is New and What is Not?" *Structural Change and Economic Dynamics* 15, no. 3: 245–264.

Chengappa, Sangeetha. 2021. More start-ups to turn Unicorns in 2021. Bengaluru, 24 April. www.thehindubusinessline.com/companies/more-start-ups-to-turn-unicorns-in-2021/article34395287.ece (accessed 11 May 2021).

Government of India. 2014. Ease of Doing Business. 25 September 2014. www.makeinindia.com/eodb (accessed 10 May 2021).

Government of India. 2015. Programme-pillars. 1 July 2015. https://digitalindia.gov.in/content/programme-pillars (accessed 9 May 2021).

Jain, Surbhi. 2016. "Growth of Startup Ecosystems in India". *International Journal of Applied Research* 2, no. 12: 152–154.

Kamat, V. Pradeep and Neela Kamat. 2020. *Education, Employability, Employment and Entrepreneurship in Relation to Five Pillars of Atmanirbhar Bharat Post COVID-19.* Mumbai, 25 July.

Massand, Ajay, M.K. Lodi and Lubna Ambreen. 2020. "Atmanirbhar Bharat: Economic, Legal, and Social Aspects". *International Journal of Management* 11, no. 6: 1038–1046.

Mesenbourg, L. Thomas. 2001. Measuring the Digital Economy. Census, Suitland: US Bureau of the Census. www.census.gov/content/dam/Census/library/working-papers/2001/econ/umdigital.pdf (accessed 9 May 2021).

Mohanta, Giridhari, Sathya Swaroop Debasish and Sudipta Kishore Nanda. 2017. "A Study on Growth and Prospect of Digital India Campaign". *Saudi Journal of Business and Management Studies* 2, no. 7: 727–731.

Nam, Chang Woon and Peter Steinhoff. 2018. "The 'Make in India' Initiative". *CESifo Forum* 19, no. 3: 44–45.

Perwez, K., Syed G. Vijayalakshmi and Syed Zahid Perwaiz. 2016. "Make in India Initiative: Reigniting Indian Manufacturing Sector". *Global Journal of Management and Business Research* 16, no. 1: 28–33.

Sarkar, Abhrajit. 2016. "Startup India: A New Paradigm for Young Entrepreneurs (A Conceptual Study)". *SSRN Electronic Journal.* https://papers.ssrn.com/sol3/papers.cfm?abstract_id=2835322

Sen, Dishani. 2020. "A Narrative Research Approach: Rural-Urban Divide in Terms of Participation in Digital Economy in India". *Journal of Management* 7, no. 1: 42–51.

Shankar, Uma. 2017. "Digital Economy in India: Challenges and Prospects". *International Journal of Research in Management Studies* 2, no. 11: 6.

Sheth, Arpan, Sriwatsan Krishnan and Arjun Upmanyu. 2021. *India Venture Capital Report 2021.* Annual, Mumbai: Bain & Company, Inc. https://ivca.in/wp-content/uploads/2021/03/bain_report_india_venture_capital_2021.pdf (accessed 10 May 2021).

Som, Lalita. 2018. "The Make in India Initiative: Has it Worked?" *Florya Chronicles of Political Economy* 4, no. 1: 55–88.

Statista. 2019. Annual growth rate of production in the manufacturing industry in India from financial year 2013 to 2019. www.statista.com/statistics/661391/manufacturing-industry-production-growth-rate-india/ (accessed 10 May 2021).

The World Bank. 2019. Ease of Doing Business in India. https://data.worldbank.org/indicator/IC.BUS.EASE.XQ?locations=IN (accessed 10 May 2021).

Usmani, Azman. 2021. Startup Street: These Could Be India's Next Unicorns. www.bloombergquint.com/business/startup-street-these-could-be-indias-next-unicorns (accessed 9 May 2021).

11 Skill Requirement in Industry 4.0

M. Laad and M. Renedo

CONTENTS

11.1 INTRODUCTION

To date, human civilisation has undergone three industrial revolutions. The first Industrial Revolution brought mechanisation, the second brought mass production and electricity, and the third brought industrial automation [1, 2]. The fourth Industrial Revolution, also known as Industry 4.0, is characterised by the application of information and communication technologies to all processes in the manufacturing industry.

Each Industrial Revolution has significantly affected entire production processes and business models including small and medium enterprises (SMEs). In each Industrial Revolution, the skills required by the workforce of various industrial organisations [3] have been greatly affected. In every transition between these industrial revolutions, few jobs disappeared while few new jobs were created. Some skills became redundant while others became more valued and desirable. The same can be expected from Industry 4.0, which is characterised by prominent technological advancements that require specially trained and skilled manpower [4].

DOI: 10.1201/9781003200857-11

Industry 4.0 has the capacity to pull individual professionals into smarter networks resulting in faster and more efficient working. In the fourth Industrial Revolution, skills that may not be considered important in today's context, will be treated as the most desired core skills in many of the professions in SMEs. This change in the required skill sets is mainly due to automation and digitalisation in the production and business processes. Future jobs will require the ability to analyse and draw meaningful inferences from generated real-time data and to make data-based decisions. With the application of automation, artificial intelligence (AI), and robots in industries, many jobs involving skills such as troubleshooting and machine problems will disappear.

In the new Industrial Revolution, production systems that already use computer technology will be further stretched into a network connection coupled with a digital twin on the internet. This digital twin will create new insights owing to data analytics and physics-based simulations in a completely virtual environment. It will facilitate the faster realisation of innovations with enhanced reliability. The digital twin will help in communicating with other facilities and generating information about same system or process, leading to automation in the production process. Networking all the systems will lead to smart factories and cyber-physical production systems, in which various systems, components, and employees will communicate via networks and the entire production process will be automated.

The rise of smart factories equipped with advanced technologies such as the internet of things (IoT), cloud computing, cyber-physical systems, multi-agent systems, artificial intelligence, and machine learning will change the current workforce from factory workers to knowledge workers in all types of enterprises. Repetitive tasks involving physical work will be automated and executed by machines and robots, while tasks requiring creativity, critical thinking, experience, intuition, and decisions-making will remain with humans. This shift in future jobs will require necessary skills and competences. Also, it will transform product manufacturing into a software-dominated business, where software development and operations will form the central part of the entire manufacturing process. As this will be a fundamental requirement, the current workforce urgently needs to gain the relevant skills and competences necessary for future jobs [5].

The digitalisation of manufacturing offers more flexibility and creates ways to provide the right information to the right people at the right time. Industry 4.0 is going to change the manufacturing, distribution, and use of products.

With the rise of Industry 4.0, the future of jobs is a major concern for governments and efforts are being made to prepare the workforce for the future with the right skills and capabilities. Disruptive technologies have intensified uncertainties around extensive job losses arising from automation and digitalisation.

Industry 4.0 carries automation through every stage of manufacturing, from processes such as procuring, processing, and raw material distribution, to the final product. Digitally reformed industries are expected to change and embrace advanced technologies such as drones, intelligent robots, sensors, actuators, artificial intelligence, machine learning, and 3D printing nanotechnology to stake a place for themselves in the ever-competitive and dynamic market.

The success of Industry 4.0 depends on the availability of relevant skills and the competences of employees in industries at every level. The specialised skills and the qualifications of workers will become important and drive the innovation and competitiveness of an organisation [6]. Organisations whose employees lack the necessary skills may face a significant drop in performance and competitiveness. Industry 4.0 will not only bring about technological advancement but also prioritise human resource development including the development of essential skills required in the future [7].

To identify the necessary skills required for Industry 4.0, it is very important to understand the changes that it anticipates bringing to the traditional manufacturing industry, the new tasks that employees will need to perform, how the tasks are going to differ from those that employees had previously undertaken, and the requirements of the new skill sets to effectively execute those tasks.

11.2 EXPECTED CHANGES IN INDUSTRY 4.0

Industry 4.0 is about establishing interconnectivity and communication between cyber-physical systems, including workpiece carriers, assembly stations, and products, people, and smart factories through the IoT. It will be necessary to develop intelligent networks throughout the value chain that can control each other through interconnected computer networks and machines. For example, machines with pre-warning systems to predict failures and activate maintenance processes automatically, or self-systematised logistics to detect and address unexpected changes in production, etc. However, this will make manufacturing and supplier networks significantly more complex.

The Industry 4.0 environment will be characterised by strong customisation of products with mass production and lots of flexibility. There will be many improvements and advancements in the area of self-automated technologies in optimisation, configuration, diagnosis, cognition, and intelligent support for employees in the growing complexity of all operations.

The fourth Industrial Revolution will drive transformation in manufacturing firms with a vision of interconnected factories with smart, data-driven, intelligent equipment. It is expected to bring significant changes to manufacturing processes with advanced-level automation and work environment interconnectivity. The machines, tools, and technologies will also change. There will be smart machines self-coordinating the manufacturing processes; service robots working together with employees on assembly lines; and advanced transportation systems for the faster and safer transfer of goods. Smart devices such as mobile phones, tablets, flexible wearables, and electronic devices will collect and analyse real-time information. The manufacturing sector is expected to make the biggest investment in automation and the internet of things. The smart factory will have a digital twin where the data generated from monitoring physical processes using virtual plant models and simulation models will be collected and analysed and these cyber-physical systems will be capable of taking timely action. In a nutshell, technology will dominate and greatly improve manufacturing and production processes.

However, the future workforce needs to be innovative, capable of deriving their own insights, and taking decisions and appropriate actions. With advanced technology using artificial intelligence and automation, industries will need employees to be equipped with different sets of skills to implement, guide, and lead this technology. Industry 4.0 is based on advancements in technology and its impact is being felt across all industries, especially SMEs.

11.3 FACTORS SHAPING INDUSTRY 4.0

According to a survey conducted by McKinsey, an exponential increase in data volumes, enhanced computational capability and connectivity, and the availability of new low power wide area networks play a very important role in shaping Industry 4.0. Increased growth in the area of data analytics and business intelligence, and advancements in technologies with human–machine interactions such as touch interfaces and augmented reality systems are also cited as important factors. Also, enhancements in the efficient transfer of digital instructions to the external world using advanced robotics and 3D printing will also be an important factor leading to the fourth Industrial Revolution.

11.4 HOW WILL FUTURE TASKS DIFFER?

There will be remarkable differences between the traditional factory and an Industry 4.0 factory. In the existing industry set-up, the focus is on offering high-end quality services and products at the lowest price and this is fundamental to growth and success. Industrial organisations strive to attain high performance output to maximise their profits and reputation. The Industry 4.0 factory will not only have self-conditioned monitoring and fault diagnosis, but also its various components and systems will be capable of self-predicting and self-diagnosis. It will offer management more insightful information and updated status on their factory. Moreover, peer-to-peer assessment and updated information on working conditions from various interconnected networks will provide a precise status report at every operation level and will compel factory management to initiate the required maintenance at the appropriate time.

Skilled workers will have a wider choice of jobs and will be associated with more than one type of job. A significant number of monotonous and ergonomically challenging jobs will disappear. Employees will collaborate with machines and share the space with intelligent robots. These robots and other machines and computer network systems will significantly support operations, though the final decisions will be made by skilled employees. At all levels, both horizontal and vertical, collaboration and teamwork will be central in the workplace.

The information and real-time data generated will gain prominence and employees will need to process data in all their day-to-day work activities and derive meaningful insights from it. Artificial intelligence and machine learning will enable tasks to be performed by both humans and machines. The human–machine interaction will be facilitated through touch, voice, and gesture.

Employees will use smartphones and tablets for communication and machine operation with ease. Employees will be required to monitor and control processes and do less physical work. New jobs will be created and will require workers to be involved in the planning process and optimisation activities.

11.5 WHAT SKILLS ARE REQUIRED IN INDUSTRY 4.0?

The World Economic Forum [8] has suggested several major changes in adapting and implementing Industry 4.0. It highlights the way that businesses view and manage all operations in existing set-ups now and in the longer term. The report also suggests that there will be a significant shift in skills requirements in Industry 4.0.

Several studies [9, 10] have reported that the skills required for Industry 4.0 are diverse as well as numerous. These skills include critical thinking, complex problem-solving, creative thinking, people management, emotional intelligence, coordination, decision-making, negotiation, and cognitive flexibility, apart from technical skills. These skills can be classified into four main categories as required by Industry 4.0:

- Technical skills
- Methodological skills
- Social skills
- Personal skills

All four skill categories are described in detail as follows:

11.5.1 Technical Skills

In the fourth Industrial Revolution, traditional factory workers of today will be transformed into knowledge workers of smart factories supported by all the enabling technologies. The fourth Industrial Revolution will witness new solutions and new technologies that can provide innovative, precise, and faster solutions.

Industry 4.0 is expected to dramatically impact the number of networking professionals who work in manufacturing and other industries. The reason for this can be termed the industrial internet of things. Industries will be transformed with many networks and there will be a huge demand for networking and IoT experts. Some examples of where networking will be required are smart factories and connected fabrication and material handling equipment, remote sensors for freight condition monitoring and inspection, automated infrastructure and smart metering for utility management and energy-saving efforts, tracking systems for vehicles and other assets, and monitoring of employee health and safety in dangerous and remote conditions.

In the automation and digitalisation of industries and processes, employees will need to be equipped with designing skills. These skills will incorporate virtualisation, simulation, interoperability, modularisation, and decentralisation capabilities, and fault and error recovery skills. Employees will also be trained to use advanced technologies such as autonomous robots and 3D printing. They will have knowledge and an understanding of process digitalisation and the ability to work with IoT-enabled systems and services.

In Industry 4.0, there will be an upsurge in real-time data gathered from customer' insights and from enterprise hardware and software, while the wheels of industry are turning rather than after the fact.

Knowledge of data science helps in making predictions and reasonable decisions based on facts. Even when it is not human beings actively drawing conclusions from all the information, there will be a huge demand for data scientists to write algorithms and build AI models. This digital revolution will create many opportunities, yet it will also be an obligation, because consumers are digitally aware, well informed, and well connected. Therefore, consumers make comparisons and seek greater value, a good experience, and satisfaction. This will require companies to technologically and culturally evolve to create their digital strategy and be innovative to carve a niche for themselves in the market.

The digitalisation of processes requires the necessary proficiency in related skills such as data analytics/data processing, IT/data/cybersecurity, cloud computing skills, IT knowledge and abilities, and artificial intelligence skills. These skills are essential for the effective use of digital information to optimise business profits and generate revenue, optimise costs, and offer desirable customer experiences.

Big data analytics will be one of the top essential skills in Industry 4.0 that employees must possess to make a significant contribution [11]. It will be more impactful if data analytics skills are coupled with technical skills, a strong business and industry knowledge, and the necessary personal and social skills [12].

As Industry 4.0 expands, programmers and software engineers with computational, simulation, and coding, computer, and software programming skills will be very much in demand. These skills will be used for designing or writing machine code for industrial control and automated systems; developing management platforms for internal industrial use and enterprise planning; creating customer-facing and mobile apps for interacting with clients on the go and doing business anywhere; data visualisation tools and dashboards for surfacing industrial data, drawing conclusions from it, and making business decisions based on it; and code to give robotic systems their marching orders and better facilitate human–machine interaction. In a nutshell, it means companies engaged with Industry 4.0 technologies need more specialists, who are well-versed in real-time operating systems and the programming languages that help build and run them, such as Python, Ruby, C, C++, and Java.

With the automation and digitalisation of processes in Industry 4.0, the risks associated with wireless and interconnected technologies would increase significantly; therefore, industries will need a robust IT infrastructure and focus on employing IT personnel. The attack surface in Industry 4.0 is large indeed. From healthcare and retail to manufacturing and supply chain management, most industries are finding useful ways to put emerging technologies to good use. Unfortunately, cyberattacks on internet-connected devices have also increased. In the coming years, all companies will have to take cybersecurity very seriously if they want to avoid losing mission-critical data or falling victim to such cyberattacks.

11.5.2 Methodological Skills

Methodological skills are skills that empower individuals to evaluate and interpret evidence, and identify and analyse situations to make a decision and solve a problem or a conflict; these are efficiency-oriented skills.

In the current Industrial Revolution, which focuses on the digital transformation of SMEs and is considered to merge three worlds (biological, digital, and physical), methodological skills could be considered one of the most important skills, since they are human features that cannot be replaced or substituted by any machine.

The structure of the labour market will change and SMEs will have to be flexible and adapt to many changes. They will have to reorganise their processes and procedures and prepare their managers and workforce for this adaptation, since humans are key to this revolution. Therefore, education and training systems should better adapt to the real market needs. Higher education organisations and vocation education and training (VET) organisations have been working on this adaptation with curricular training programmes, but there is still much work to be done. The integration of education and training programmes and SMEs and their reality is vital. Nowadays, a very popular trend is to learn by doing, that is, to simulate real learning environments and experience real situations, also known as workplace learning.

Since the start of the Industry 4.0 revolution, workforces have become increasingly dispensable and many jobs have disappeared, but some new ones have been created. The labour market in this sense is changing and so is the change to job demand; however, several skills cannot be replaced by technology (yet!), so a more skilled and competent workforce with expertise and also more qualified managers are in demand.

Creativity and innovation could be considered the main methodological skills to master in this future labour framework. The more complex a process is, the more preparation a human requires, and the main key competences for this training are creativity and innovation. "Innovate or die" [13] became the goal of companies over the past two decades, and innovation was crucial to succeed when globalisation started to emerge.

Essentially, innovation is a process of turning opportunities into new ideas and putting these ideas into practice [14], so the ability of a company to grow will depend on its ability to be creative and innovative.

There are different definitions of creativity, such as the one concreted by Flynn (2003) [15], who divided creativity into three categories: normative creativity (focused on generating ideas to solve specific needs, problems, and objectives), explorative creativity (focused on generating a broad spectrum of ideas, not necessarily related to any need), and creativity by serendipity (focused on the innovative idea being discovered by accident); or the definition stated by Amabile (1998) [16], who divided it into three different components: expertise (technical and intellectual knowledge of an individual), creative thinking (applying an individual's skills to imaginative problem-solving), and motivation (intrinsic and extrinsic factors influencing an individual's creativity).

However, it can be stated that these skills are absolutely human and they will be essential in future jobs, so creativity and innovation are vital skills to teach and train workforces and leaders, since they are the people who can sensitise what does and what does not work. Humans are the only ones capable of thinking out of the box and pondering efficient and unique ideas.

A key skill closely related to innovation is entrepreneurial thinking. It was already proved that "innovative entrepreneurs spend 50% more time on discovery activities (questioning, observing, experimenting and networking) than others with no innovation track record" [17]. However, they do not work to gain promotion, but to offer diverse ideas and perspectives in order to contribute to their organisation's development and progression.

Mostly linked to managers, entrepreneurial thinking has much to do with inspiring and involving the workforce and with boosting their curiosity and creativity, connecting them, providing them with the knowledge, abilities, and attitudes necessary for their future jobs and tasks, and even strengthening the links between the educational world and the real labour world and promoting workforce inclusion and growth.

Undoubtedly, emotional intelligence can be considered the core skill of the methodological competences and it can influence different social and personal competences and skills, such as intercultural, networking, and teamwork skills.

The concept of "emotional intelligence" was first described by Salovey and Mayer [18] as "the ability to understand how others' emotions work and to control one's own emotions", and later widened by Goleman to include such competencies as optimism, conscientiousness, motivation, empathy, and social competence [19].

It is a fact that humans cannot compete with machines as regards knowledge storage, but they have unique features and abilities such as attitude, self-motivation, self-esteem, self-development, self-knowledge, self-awareness, self-regulation, and creative thinking through feeling and thinking processes. Individuals can learn to be open to the new, to be aware of the barriers to expressing creativity, and they can gain self-confidence, awareness and amplitude, mental flexibility, etc. And the most important thing is that all of this can be trained and improved.

To finish with the most important skills for the workforce to acquire in order to become highly skilled and be ready for the new labour market requirements, it is necessary to mention decision-making and problem-solving, directly related to the emotional intelligence skill.

Issues that arise need to be discussed and solved by working through the details. In this way, creativity and productivity are increased, communication is improved, and team member bonds are strengthened. Some mandatory steps to solving a problem are to define it and find its nature and causes; generate new ideas and brainstorm (if there is the possibility to inspire other team members to do so and share ideas, so much the better); evaluate and select a solution; and implement and evaluate the outcomes after its implementation.

Decision-making is the process of making choices by gathering information and evaluating alternative solutions [20]. In this process, there are seven steps to organise all the information before making a decision: identify the decision and its nature; collect relevant information; identify the alternatives; weigh the evidence, imagine the decision implementation, and evaluate its suitability; choose among all the alternatives or a combination of some of them; implement the decision; and review it and its consequences.

Taking into account all these competences, it is evident that upskilling and reskilling the workforce are indispensable to educational and training providers and to SMEs in order to employ highly skilled individuals ready to face the challenges and changes that the fourth Industrial Revolution is provoking in the labour market.

11.5.3 Social Skills

Social skills are the tools that enable employees to communicate, learn, seek help, and develop healthy relationships in the workplace. Employees foster and maintain meaningful relationships with employers, colleagues, and network contacts with good social skills. Social skills are beneficial for professional growth and success. A significant number of studies [21, 22] have reported that, although technical skills are essential, they must be complemented by strong social and personal skills. Social skills include intercultural skills, communication skills, language skills, networking skills, the ability to work in a team, the ability to compromise and be cooperative, and the ability to transfer knowledge and leadership skills.

The future's fast-moving, technologically advanced, and well-connected world will essentially need employees to adapt to the new actualities of Industry 4.0 if they wish to be successful.

There will be rapid change in the future workplace and employees are expected to be agile and ready to embrace the change. This change should be considered as an opportunity to grow and innovate.

Future employees will be required to be emotionally intelligent. They should be able to perceive, recognise, and manage their emotions as well as those of their team members and colleagues and confidently take decisions. Tomorrow's workforce will need to have confidence in their own abilities and decisions, yet be humble about their achievement and position in the organisation.

Employees who acquire all the essential skills will rise to leadership positions and will take the initiatives, drive the business, and help others to grow and shine. The work environment of Industry 4.0 will be more transparent and collaborative and it will be crucial that the employee's actions are aligned with the organisation's goals and objectives.

Employees in Industry 4.0 will need to understand the roles, responsibilities, expectations, and aspirations of each stakeholder in the organisation. They will need to have a big-picture vision and a deep understanding of the impact of their decisions on each stakeholder. It will be important to strategically identify and meet the requirements of each stakeholder effectively.

The future work environment will be collaborative in nature. It will be imperative that employees get to know each other's strengths and weaknesses and strive to acquire new skills and grow professionally.

Employees of Industry 4.0 need to be culturally intelligent as the future workplaces will be very diverse and globalised. Employees will need to adopt, appreciate, and leverage the cultural differences that every member brings to the team, as well as their intergenerational differences, which can provide crucial knowledge and expertise too.

Further social skills are negotiation, perspective taking, professional ethics, understanding diversity, self-awareness, self-organisation, and interpersonal skills. In Industry 4.0, it will be important to arrive at common agreements for the mutual benefit of everyone in the organisation. This will not only create a more collaborative and participative environment, but also encourage and stimulate employees to experiment and contribute new ideas. It will also help in building a learning and innovative environment promoting decision-making and developing a more open and digital mindset. Social perception will also be a highly desirable skill in Industry 4.0, especially for future leaders to know how to motivate people to function in a digital environment, ensuring their whole-hearted participation and developing an atmosphere of trust and cooperation.

11.5.4 Personal Skills

Personal skills refer to the ability to act in a reflective and autonomous way in any given situation. People with personal skills are flexible, tolerant of ambiguity, motivated to learn new things, do not lose control or panic under pressure, and have a positive and sustainable mindset and compliance. These skills can be used in different situations and work environments. Personal skills are not specific to a particular job or assignment or academic discipline, but are applicable to a wide variety of tasks, conditions, and work environments.

Future employees will need to be flexible to accommodate the changing demands of people, workspaces, and tools, and be willing to embrace the new reality.

The people from different age groups and different cultural backgrounds, future employees will need to show authenticity. They should have the ability to build strong connections based on mutual trust. Employees need to exude the same confidence and authenticity in situations of uncertainty and failure as when they are successful and growing.

The future workplace will be dynamic and bring about many changes at a very fast pace. Employees will need to adapt quickly to new situations and yet maintain focus on the goals and objectives of the organisation. They need to be stable and focused even when the pace of change could compel them to pivot from one priority to another. An employee would be expected to be able to cut through the chaos and contribute towards the growth of the organisation. A quick assessment of a situation, fast decision-making, and taking appropriate action would be an essential requirement for future employees. The following are some of the essential personal skills that employees need to acquire in Industry 4.0.

With the greater influx of machines and automation in business in future workplaces, employees will need to be even more intuitive. They will need to develop a unique human skill, i.e. the ability to "read" what's not being said, to perceive and understand human emotions. A person who is willing to have conversations around what's not being said and has the patience and tolerance to hear the good as well as the bad from their colleagues will be poised for future success as a leader.

The future workplace will change at a very fast pace. There will be no clear roadmap or instructions to follow. Employees will need to have the courage to face the challenges and navigate unknown situations and circumstances. It will be essential for future employees to have the courage to change a course of action if a situation calls for a new strategy.

Future employees will also need to be open to training and learning to upgrade their knowledge and skills with the ever-changing technology. They will have to develop an attitude for lifelong learning. Lifelong learning should not just be seen as a consequence of the new industrial environment, but it should also be considered in demographic terms, acknowledging the increased life expectancy of people with prolonged working lives, in contextual terms; and also in light of ease of access to information and knowledge. In addition to rethinking current education plans, future employees will need to continuously upgrade their knowledge and skills by continuous learning and reskilling.

11.6 CONCLUSION

Industry 4.0 is expected to start a trend of automation and data exchange in manufacturing industries. The growth of Industry 4.0 and smart factories together with all the enabling technologies will transform traditional factory workers into knowledge workers. There are also various challenges to implementing Industry 4.0 in terms of skills requirements and manpower with required skills. Many tasks involving physical and routine work will be executed by machines or robots. Activities requiring creative thinking, critical thinking, intuition, experience, and decision-making will remain with humans.

Technological skills, methodological skills, social skills, and personal skills will be very relevant in Industry 4.0. Future employees will need to explore different technologies. They will need to have strong digital skills, programming skills, and design skills and a thorough understanding of the operation of the machines, tools, and technologies used in process design, ergonomics, etc. In the future, employees will also need to address issues of a sustainable environment at the very early stages in a product's design [23].

Methodological skills will be highly desirable in Industry 4.0. These skills include creativity, problem-solving, entrepreneurial thinking, decision-making, and analytical skills. Methodological skills will be one of the top skills that employees will need to develop. Critical thinking skills will allow employees to think logically to identify the strengths and weaknesses of decisions, possible solutions, and conclusions or approaches. This will also help in understanding the implications of the available information for existing as well as future problem-solving and decision-making.

Social skills such as negotiation, perspective taking, professional ethics, understanding diversity, self-awareness, self-organisation, and interpersonal skills will be essential skills for future employees to acquire. Employees will need to develop strong professional relationships for the common benefit and work in a collaborative and participative manner. It will be important to create an atmosphere that will stimulate employees to experiment and contribute with innovative ideas, encourage decision-making, and promote an open and digital mindset.

Personal skills include cognitive flexibility, ability to work under pressure, motivation to learn, and a sustainable mindset and compliance. These skills will be in high demand and highly desirable. Logical reasoning, problem sensitivity, mathematical reasoning, and visualisation will also be considered essential skills. Written and verbal communication skills, digital/information and communication technology, computer literacy, numeracy, people management skills, and evaluation skills will be some of the most desired skills in Industry 4.0 for the successful implementation of and adaptation to Industry 4.0.

However, in Industry 4.0, advanced technologies will not replace humans for higher productivity; rather, there will be close human–machine collaboration at work. With rapid changes in the nature of work, future employees will need to constantly upgrade their knowledge and relevant skills. They will need to develop an attitude and a willingness for lifelong learning to meet the challenge of essential skills requirements in Industry 4.0. Interdisciplinary skills development will be necessary to ascertain the effectiveness and productivity of employees in an organisation.

REFERENCES

1. Schwab, K. “The fourth industrial revolution”. *Encyclopedia Britannica*, 23 March, 2021. www.britannica.com/topic/The-Fourth-Industrial-Revolution-2119734.
2. Darwish, Hasan. “Expanding industrial thinking by formalizing the industrial engineering identity for the knowledge era”. PhD diss., North-West University, 2018.
3. Maisiri, Whisper, Hasan Darwish, and Liezl van Dyk. “An investigation of industry 4.0 skills requirements”. *South African Journal of Industrial Engineering* 30, no. 3 (2019): 90–105.
4. Benešová, Andrea, and Jiří Tupa. “Requirements for education and qualification of people in Industry 4.0”. *Procedia Manufacturing* 11 (2017): 2195–2202.
5. Fitsilis, Panos, Paraskevi Tsoutsa, and Vassilis Gerogiannis. “Industry 4.0: Required personnel competences”. *Industry 4.0* 3, no. 3 (2018): 130–133.
6. Mavrikios, Dimitris, Konstantinos Georgoulias, and George Chryssolouris. “The teaching factory paradigm: Developments and outlook”. *Procedia Manufacturing* 23 (2018): 1–6.
7. Schallock, Burkhard, Christoffer Rybski, Roland Jochem, and Holger Kohl. “Learning factory for Industry 4.0 to provide future skills beyond technical training”. *Procedia Manufacturing* 23 (2018): 27–32.
8. World Economic Forum. “The future of jobs: Employment, skills and workforce strategy for the fourth industrial revolution”. Global Challenge Insight Report (2016). Available at http://reports.weforum.org/future-of-jobs-2016/
9. Hecklau, Fabian, Mila Galeitzke, Sebastian Flachs, and Holger Kohl. “Holistic approach for human resource management in Industry 4.0”. *Procedia Cirp* 54 (2016): 1–6.
10. Leinweber, Stefan. “Etappe 3: Kompetenzmanagement”. In *Strategische Personalentwicklung*, pp. 144–178. Springer, Berlin, Heidelberg, 2010.
11. Anon. *The digital revolution: The impact of the fourth industrial revolution on employment and education*. Edge Foundation, London, 2016. http://www.edge.co.uk/media/193777/digital_revolution_web_version.pdf.

12. Venkatraman, Sitalakshmi, Tony de Souza-Daw, and Samuel Kaspi. "Improving employment outcomes of career and technical education students". *Higher Education, Skills and Work-Based Learning* (2018).
13. Getz, Isaac, and Alan G. Robinson. "Innovate or die: Is that a fact?". *Creativity and Innovation Management* 12, no. 3 (2003): 130–136.
14. Tidd, Joe, and John R. Bessant. *Managing innovation: Integrating technological, market and organizational change.* John Wiley & Sons, Hoboken, NJ, 2020.
15. Flynn, Michael, Lawrence Dooley, D. O'Sullivan, and K. Cormican. "Idea management for organisational innovation". *International Journal of Innovation Management* 7, no. 4 (2003): 417–442.
16. Amabile, Teresa M. *How to kill creativity.* Vol. 87. Harvard Business School Publishing, Boston, MA, 1998.
17. Dyer, Jeff, and Hal Gregersen. "Learn how to think different(ly)". *HBR Blog Network*, 27 September (2011).
18. Salovey, Peter, and John D. Mayer. "Emotional intelligence". *Imagination, Cognition and Personality* 9, no. 3 (1990): 185–211.
19. Goleman, Daniel. *Emotional intelligence.* Bantam, New York, 2005.
20. Burduk, Anna, Edward Chlebus, Tomasz Nowakowski, and Agnieszka Tubis, eds. *Intelligent systems in production engineering and maintenance.* Vol. 835. Springer, 2018.
21. World Economic Forum. "The future of jobs: Employment, skills and workforce strategy for the fourth industrial revolution". Global Challenge Insight Report (2016).
22. Selamat, Ali, Rose Alinda Alias, Syed Norris Hikmi, Marlia Puteh, and S. M. Tapsi. "Higher education 4.0: Current status and readiness in meeting the fourth industrial revolution challenges". *Redesigning Higher Education towards Industry* 4 (2017): 23–24.
23. Grajewski, Damian, Jacek Diakun, Radosław Wichniarek, Ewa Dostatni, Paweł Buń, Filip Górski, and Anna Karwasz. "Improving the skills and knowledge of future designers in the field of ecodesign using virtual reality technologies". *Procedia Computer Science* 75 (2015): 348–358. https://doi.org/10.1016/j.procs.2015.12.257.

ANNEXURE A

Other sources

Abraham, A., The Need for the Integration of Emotional Intelligence Skills in Business Education, 2006. https://ro.uow.edu.au/commpapers/238

World Economic Forum, The Future of Jobs Employment, Skills and Workforce Strategy for the Fourth Industrial Revolution, 2016. Available at http://reports.weforum.org/future-of-jobs-2016/

World Economic Forum, The Future of Jobs, World Economic Forum, 1-167, 2016.

INDUSTRY 4.0: REQUIRED PERSONNEL COMPETENCES

Panos Fitsilis, Paraskevi Tsoutsa, Vassilis Gerogiannis

Faculty of Business and Economics, University of Applied Sciences of Thessaly, Greece

Department of Mathematics, University of Patras, Greece

(2018)

TRANSFORMATION OF THE MACHINES FROM LEARNER TO DECISION MAKER: INDUSTRY 4.0 AND BIG DATA

Şebnem ÖZDEMİR, Department of Management Information System, Beykent University, Turkey, sebnemozdemir@beykent.edu.tr
Enhancing Decisions and Decision-Making Processes through the Application of Emotional Intelligence Skills
James D. Hess
Department of Family Medicine,
Oklahoma State University Center for Health Sciences,
Tulsa, Oklahoma, USA
and
Arnold C. Bacigalupo
Voyageur One, LaGrange, Illinois, USA
Key Competencies for Industry 4.0
Katarzyna Grzybowska1,* and Anna Łupicka2 (2017)
Holistic Approach for Human Resource Management in Industry 4.0
Fabian Hecklau, Mila Galeitzke, Sebastian Flachs, Holger Kohlb (2016)
Scientific Mapping to Identify Competencies Required by Industry 4.0
Liane Mahlmann Kipper, Sandra Iepsen, Ana Julia Dal Forno, Rejane Frozza, Leonardo Furstenau, Jessica Agnes, Danielli Cossul (2021)
The Industry 4.0 Induced Agility and New Skills in Clusters
Marta Götz (2019)
Schumpeter (1934)
Annabeth Aagaard (2011)
The Innovators DNA (2011)
(What Is Problem Solving? Steps, Process & Techniques | ASQ, n.d.).

12 The Changing Role of Academics from the Perspective of Educational Transformation in Education 4.0

F. Karaferye

CONTENTS

12.1 INTRODUCTION: BACKGROUND AND DRIVING FORCES

The fourth Industrial Revolution with its new technologies and futuristic innovations, from artificial intelligence (AI) to humanoid robots, has started to bring about dramatic changes to human life; some examples are shown in Table 12.1. These dramatic changes include changes to business models in all sectors, and the increasing pace of change in jobs (creation, destruction, new forms) with the realignment of the workforce in the education and industrial sectors, while expected new knowledge, skills, and competencies of individuals are increasing (Echeberria, 2020; World Bank, 2019; World Economic Forum [WEF], 2018a, 2018b).

Emerging in-demand roles come in parallel with those emerging changes listed above, such as new roles and positions as data scientists, software developers, and e-commerce specialists, i.e. roles that are based on science, technology, engineering, and mathematics (STEM) (WEF, 2018a). In addition, these emerging changes also highlight the need for distinctive "human skills". To illustrate the roles emphasising "human skills" or "soft skills", roles and positions in sales and marketing and training and development are examples (WEF, 2018a). Additionally, every individual independent of their field of work/occupation will need to develop multiple literacies

DOI: 10.1201/9781003200857-12

TABLE 12.1
Emerging Changes in the Fourth Industrial Revolution

Technological advances

- High-speed mobile internet
- Artificial intelligence
- Big data (analytics)
- Cloud technology
- Machine learning
- Augmented and virtual reality
- Robotics technologies (stationary robots, non-humanoid land robots, aerial drones, underwater robots, etc.)
- The internet of things and cyber-physical systems

Social and economic changes

- National economic growth trajectories
- Expansion and modernisation of education
- Greener global economy through advances in new energy technologies
- Changing geography of production
- Changing distribution and value chains

Changes in employment and tasks

- Productivity-enhancing roles
- Remote staffing beyond physical offices and decentralisation of operations
- Human–machine frontier within existing tasks
- Division of labour and collaboration between humans and machines in new roles and tasks

Source: Adapted from WEF (2018a).

throughout their lives for full participation in the knowledge-based economies and societies of the 21st century with those changes stated above. At the fundamental level, those can be listed as literacy, numeracy, and problem-solving in today's technology-rich environments (Organisation for Economic Co-operation and Development [OECD], 2019). Several of the lifelong literacies needed in this rapidly changing era are provided, as the Framework for 21st Century Learning (2019) states:

- Global awareness
- Financial, economic, business, and entrepreneurial literacy
- Civic literacy
- Health literacy
- Environmental literacy

From a socio-economic perspective, the labour market is continuously being affected by these deep and rapid transformations in addition to globalisation and demographic changes. "Employers are demanding new skills and qualified workers, while many people are looking for a job" including new graduates (OECD, 2019). Additionally, reports state that there is a "skills gap", a gap between the knowledge and skills of new graduates and what employers need in the workplace. The gap is also an issue for the workforce in the workplace, which is why more and more organisations are investing in reskilling and upskilling their employees.

Since the job market will continue to shift in response to new technologies, which highlight the skills needed for future jobs and the concept of lifelong learning with the increasing need for upskilling, reskilling, and other broader learning needs, investment in closing the gap will continue. Furthermore, the Covid-19 pandemic and the ongoing global recession have also added to the transformation of the global jobs and skills landscape. Consequently, the need for reskilling, upskilling, and learning in a broader aspect has accelerated (World Economic Forum [WEF], 2021a). According to WEF's *Future of Jobs Report* (2020), 50% of all employees will need reskilling in the next five years while 40% of current workers' core skills are expected to change (WEF, 2020a).

Meanwhile, as the reports state, there is a disconnection between education systems and labour markets, which when added to the rapid technological advancements, eventually adds to the increasing global skills gap (WEF, 2020b). On this, the project *Closing the Skills Gap* 2020 is illustrative of a sample project that set out to develop a business commitment framework to drive strategic action among global companies to invest in upskilling and reskilling their workforces. Organisations were invited to design new initiatives and/or align existing education and training initiatives, addressing at least one of the target areas shown in Figure 12.1.

Based on the relevant research, it is clear that a holistic resolution approach is needed to manage the growing global skills gap. Thus, governments, education institutions, and business organisations from different sectors need to collaborate

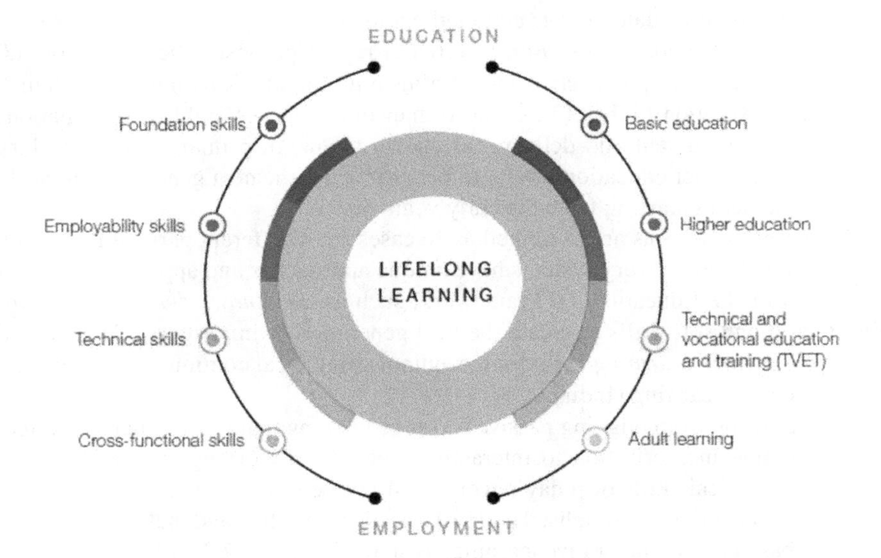

FIGURE 12.1 Closing the Skills Gap 2020 Business Commitment Framework. Source: Taken from WEF (2020b) *Closing the Skills Gap: Key Insights and Success Metrics*, p. 6.

and take steps towards innovative and sustainable solutions. In the above-mentioned project, with this purpose, organisations were invited to design new or align existing education and training initiatives addressing basic education, higher education, technical and vocational education and training, and adult learning, which were developed with a lifelong learning approach as parts of education and training ecosystems.

Consequently, since those changes demand a response from educational institutions, the question arises: Are educational institutions ready, well-equipped, and well-prepared to meet the needs and lead the change towards the future? Following this question, the chapter discusses the educational needs and the concept of Education 4.0, and the changing role of academics from the perspective of educational transformation in this era and beyond.

12.2 THE FOURTH INDUSTRIAL REVOLUTION AND EDUCATION: EDUCATION 4.0

Today, not only are education systems facing increasing demands all over the world but schools are also facing increasing demands to prepare students for rapid economic, social, technological, and environmental changes: "for jobs that have not yet been created, for technologies that have not yet been invented, and to solve social problems that have not yet been anticipated" (OECD, 2018). Furthermore, faced with unprecedented social, economic, technological, and environmental challenges, education still plays a critical role in an individual's life and society, now more than ever. With quality education, individuals can prepare for life in this era; they can shape their own lives and further contribute to the well-being of others and the global community. This critical role requires education systems and schools to adopt more responsive and up-to-date models and methodologies.

The WEF's *Schools of the Future* (2020c) report proposes the Education 4.0 Framework from the perspective of defining high-quality learning in the fourth Industrial Revolution (Table 12.2). The framework proposes eight transformations within learning content and delivery, as shown below, in primary and secondary schools – pre-higher education level – to better prepare the next generation of workforce and societies starting from the early years on.

In the report, schools are presented as 16 cases from different parts of the world, whose content and delivery systems have unique approaches and apply to a different dimension of the Education 4.0 Framework, such as *the Knowledge Society*: combining hard and soft skills to create the next generation of innovators (Canada) and *Pratham's Hybrid Learning Programme*: empowering local communities to support student-centred learning (India).

The transition from ongoing passive forms of learning that mostly focus on direct instruction and memorisation to interactive methods of learning that promote the highlighted critical skills of today's world will not be quick. Yet, creating learning ecosystems that allow personalised and self-paced, accessible, and inclusive project-/problem-based and collaborative learning, as stated in Table 12.2, will pave the way.

Another future projection on education and school systems is provided by the OECD (2020) in *Back to the Future of Education: Four OECD Scenarios for*

TABLE 12.2
WEF Education 4.0 Framework

	Areas	Brief information
1.	Global citizenship skills	Includes content that focuses on building awareness about the wider world, sustainability, and playing an active role in the global community.
2.	Innovation and creativity skills	Includes content that fosters skills required for innovation, including complex problem-solving, analytical thinking, creativity, and systems analysis.
3.	Technology skills	Includes content that is based on developing digital skills, including programming, digital responsibility, and the use of technology.
4.	Interpersonal skills	Includes content that focuses on interpersonal emotional intelligence, including empathy, cooperation, negotiation, leadership, and social awareness.
5.	Personalised and self-paced learning	Move from a system where learning is standardised, to a system based on the diverse individual needs of each learner, and is flexible enough to enable each learner to progress at their own pace.
6.	Accessible and inclusive learning	Move from a system where learning is confined to those with access to school buildings to a system in which everyone has access to learning and is therefore inclusive.
7.	Problem-based and collaborative learning	Move from process-based to project- and problem-based content delivery, requiring peer collaboration and more closely mirroring the future of work.
8.	Lifelong and student-driven learning	Move from a system where learning and skilling decrease over one's lifespan to a system where everyone continuously improves on their existing skills and acquires new ones based on their own individual needs.

Source: Adapted from WEF (2020c, p. 4).

Schooling, Educational Research, and Innovation; the four scenarios are presented in the book as shown in Table 12.3.

The four scenarios were developed with the purpose to consider probable and improbable changes in the future. These can help identify challenges and opportunities in advance to better prepare for the future and start acting today. It is certain that education will and "must evolve to continue to deliver on its mission of supporting individuals to develop as persons, citizens and professionals" in this complex and rapidly changing world. Of importance too is that such changes are to apply to higher education, adult education, and, in short, lifelong learning following basic education as the cornerstone of the whole education ecosystem (OECD, 2020).

12.3 HIGHER EDUCATION AND THE CHANGING ROLE OF ACADEMICS IN EDUCATION 4.0

The current conceptualisation of education in higher education institutions is still largely rooted in the model of Industry 3.0 despite the use of cloud computing,

TABLE 12.3
The Four OECD Scenarios for the Future of Schooling

Scenario 1	Schooling extended	Participation in formal education continues to expand. The structures and processes of schooling remain. More individualised learning is supported.
Scenario 2	Education outsourced	Learning takes place through more diverse, privatised, and flexible arrangements. Traditional schooling systems break down.
Scenario 3	Schools as learning hubs	Schools open to their communities favouring forms of learning, civic engagement, and social innovation.
Scenario 4	Learn as you go	Education takes place anywhere at any time. There is no longer a distinction between formal and informal learning.

Source: Adapted from www.oecd-ilibrary.org/sites/178ef527-en/index.html?itemId=/content/publication/178ef527-en#component-d1e153.

learning analytics, big data, and artificial intelligence in teaching and learning environments – to different extents in different institutions (Cheung et al., 2021; United Nations Educational, Scientific and Cultural Organization [UNESCO], 2014).

Over the years, but especially after the 1980s, the role of higher education institutions quickly broadened. In a report at the beginning of the 2000s before the expanding research on the 21st-century education framework or before the fourth Industrial Revolution was mentioned, some similar challenges and needs to those of today were put forward upon listing the goals of the higher education area (OECD, 2004):

- Upskilling and lifelong learning
- Social inclusion, widening participation, citizenship skills
- Economic development
- National/regional policies
- Cultural development and regeneration
- Knowledge-based developments
- Research and development, especially in science and technology

Today, within the 21st-century education framework and the effects of the fourth Industrial Revolution following technological/digital transformations, mobilisation, academic internationalisation, and globalisation, more challenges and needs have been added to the list (Bonfield et al., 2020; Karaferye, 2020):

- How to minimise the skills gap between higher education and the labour market.
- How to increase student experience and skills – bridging between education and work life.
- Continuous professional development for faculty, management, and staff.
- Setting up a digital infrastructure.

- How to bring and adapt high tech innovations into teaching and learning environments.
- Managing demographics/increasing student numbers.
- Internationalisation and global competition.
- Aligning with and supporting sustainable development goals.
- Supporting well-being and a positive climate.

Here, another challenge that higher education institutions and other sectors globally are facing today is the effects of the Covid-19 pandemic, which began in the last quarter of 2019 and is still a threat in 2021. Distance education and/or online education had already been an option or an opportunity provided by institutions. However, with the pandemic, whether or not they had the necessary infrastructure, higher education institutions had to shift their teaching and learning processes online in an unprecedentedly short time.

Even though it was necessary to continue education through digitalisation, it led to new challenges and requirements, such as training and support for staff, basic digital infrastructure, and so on. As it is seen in the literature, after the quick shift from face-to-face to online education, relating to from physical and digital infrastructure needs to mental, cognitive, social and emotional needs, both on the academics' side and students' side negative experiences have also been reported in the findings (Fernández-Cruz et al., 2020; Sahu, 2020).

Even before the pandemic, with the era's technological advances, there have certainly been some digital trends in education and in higher education institutions. Yet, with the quick switch to distance/online learning, concerns over the effectiveness of existing organisational support for university staff (academics, management, and administrative staff) and academics' adoption of/adaptation to digitalisation in education, emerging needs and expectations have increased (Raghunath, Anker, & Nortcliffe, 2018; Zhou & Milecka-Forrest, 2021).

The study by Zhou and Milecka-Forrest (2021) on digitalisation in education, examined the motivations of and the constraints on academics when adopting digital technologies. In the study, the main motivations of academics in using digital technologies in teaching and learning processes included to improve students' learning, collaboration and community building, effective and efficient use of resources, and staff development. However, it is proposed that the study's academics had concerns over the adoption of digital technologies, referring to discouragement regarding student learning experiences, technical capabilities, workload, and staff development.

To manage such common discouragement and constraints, merely improving the digital infrastructure and providing staff and students with basic literacy on the use of digital tools and platforms would not yield satisfactory results. In fact, a variety of continuous professional development opportunities are needed for academics, such as seminars, workshops, online/offline courses, professional learning communities (PLCs) for exchanges, mentoring, and coaching, following needs analyses and gathering the expectations of staff in addition to other institutional capacity building agendas. Since digitalisation and technological advances will continue to influence teaching and learning content and delivery mechanisms, investing in the competencies of today's labour force is a must.

When the current situation in education is considered, the impact of Education 4.0 and, to some extent, the push of the pandemic caused an unprecedented quick shift to digital/smart education environments. Consequently, use of the elements of the fourth Industrial Revolution (as listed below) was hastened and heightened in higher education, to different extents in different higher education institutions (even differences among departments) (Cheung et al., 2021; Matthews, McLinden, & Greenway, 2021; Cevik-Onar et al., 2018; Whalley et al., 2021; Xing & Marwala, 2017; Xu, Xu, & Li, 2018):

- Artificial intelligence
- Augmented reality (AR)/virtual reality (VR)
- Massive open online courses (MOOCs)
- Cloud computing
- Machine learning
- Robotics
- Big data and analytics
- Learning analytics
- Gamification
- Cyber-physical systems
- The internet of things
- The internet and social media
- Virtual classrooms and labs, virtual libraries, and virtual teachers
- The use of mobile devices
- Blended learning

These elements have brought both challenges and opportunities to the teaching and learning environments. Some challenges are changes in the roles of academics, the student's role, and a paradigm shift from centralised to decentralised processes while integrating formal and informal education. On the other side, some opportunities could be exemplified as accessibility, flexibility, and variety through smart learning environments while also allowing personalised learning (Whalley et al., 2021).

The change in roles could still be quite confusing due to the rapid transition from the traditional understanding of lectures where students assemble and mostly listen to a lecturer, to synchronous/asynchronous virtual classes. Now, there are higher education institutions where academics are transferring old habits of face-to-face processes to online processes as much as possible, where a lecturer talks online and leaves online assignments or administers online tests. Academics in some institutions have built smart learning environments with specifically designed instructional resources, using pedagogical approaches that pay attention to psychosocial aspects, personalised adaptive learning, and so on, instead of the one-size-fits-all learning activities. In addition, assessment processes have changed to more formative processes (Cheung et al., 2021). These differences among higher education institutions create differences among the roles of academics.

What next? Digital transformations and technological advances will continue to evolve. Using the era's new technologies such as artificial intelligence, augmented/virtual reality, and machine learning will probably continue, presenting new learning tools and services (Bonfield et al., 2020). Good practices using digital technologies in higher education were already in existence before the rapid move of the last two years. In a study by Bonfield et al. (2020), those practices are exemplified as four cases projecting into the future as different scenarios in higher education. These implementations of digital technologies in higher education institutions are shown in Table 12.4.

Lessons to learn from these practices could be that firstly, in the transformation of learning environments into smart ones together with making the transition between

TABLE 12.4
The Four Scenarios for Higher Education Institutions

Scenario 1	Deakin University	Smart Campus Programme	Launched in 2015, aimed at using cutting-edge digital technologies to digitise physical campus environments to provide campus users with a smart, personalised, responsive, and enriched campus experience.
Scenario 2	Deakin University	Deakin Genie Digital Assistant	In 2017, Deakin introduced Genie – a smart digital personal assistant – to help students during their life at the university. It can interrogate any data source including student records, timetables, financial data, course and unit information, assessment data, and library information.
Scenario 3	Deakin University/ University of Bath	Online Postgraduate Degrees/ Individual MOOCs	This project began in 2017 with a view to launching eight postgraduate degrees on the FutureLearn platform. It adopted a holistic approach to design work in digital education within the Degree Design Thinking Framework with the key elements: portfolio design, learning design, service design, and team design.
Scenario 4.	Nanyang Technological University (NTU), Singapore	Lifelong Learning and SkillsFuture courses	Within the national movement to upskill and train Singaporeans of all ages, a Training and Adult Education Industry Transformation Map in collaboration with unions and industry was launched. Professional and continuing education programmes aligning with the map included short/semester-long courses, executive programmes, mobile courses, and courses as part of a part-time engineering degree programme at NTU.

Source: Bonfield et al. (2020, p. 230).

education and workplace more connected and smoother, collaboration among stakeholders – governments, business organisations, and higher education institutions – is fundamental. Secondly, any transformation needs time and effort with careful planning and the involvement of all actors from the very first step. Another observation is that words related to "digital", "smart", and "technology" are used repeatedly. However, human-related skills – soft skills – such as interpersonal skills are equally important. In other words, in addition to the increased training in digital skills and science, technology, engineering, (arts), and mathematics (STE[A]M), training in soft skills is also highly valued (PwC, 2018).

For instance, in a study by Cevik-Onar et al. (2018), because departments of engineering prepare the future workforce who will be working with the technologies of cloud computing, data analytics, artificial intelligence, machine learning, adaptive robotics, and so on, these technologies should also be included and taught comprehensively in engineering departments. Additionally, soft skills are as important as the hard skills mentioned, because today's students will most probably be working in interdisciplinary teams in their future workplaces. This means that before they graduate from higher education, they need to find multiple opportunities to work collaboratively with others to fulfil interdisciplinary tasks in teams of students/stakeholders with different roles/from different fields. In contrast, from the traditional perspective, students of a single engineering field of study (i.e. industrial engineering) focus only on that field's tasks and responsibilities and mostly tech-based skills. However, it is not easy to later equip those students with the necessary hard skills, soft skills, and interdisciplinary skills required for Industry 4.0 at the workplace (Cevik-Onar et al., 2018), and this is the root of skills gap problems. Thus, departmental/field of study programmes needs to be modernised to better prepare graduates for the future, considering both hard skills and soft skills. Here to continue the same example, students from different engineering departments could be assigned interdisciplinary tasks – employing both hard skills of different disciplines and soft skills to work effectively and efficiently with others. Moreover, it would be even more fruitful to assign students from different faculties to create and allocate more real-life projects/problem-based tasks, roles, and responsibilities to students. The WEF (2021b) report provides another future projection on higher education and lifelong learning with the use of digital technologies (Figure 12.2).

In the education longitudinal emergence scatter plot (LEnS) provided, two educational norms for the future are projected. The first norm, "institutional credentials", is something similar to today's trends. In those, structured education systems and universities will continue, but they will require academics to employ new tools and teaching methodologies addressing the needs and expectations of the time. The second norm, "self-taught skills", presents a totally different scenario compared to today's trends. In this possibility, the competencies of the individual, real-life problem-solving skills, and value creation (meritocracy) are prized rather than an institution/school-based delivered credits/accreditation. As can be seen, in a probable future, with the effect of growing lifelong learning needs and university tuition costs, nano degrees (online learning credential programmes on any topic) are expected to increase, too. Also, digitised and virtualised content and delivery mechanisms are

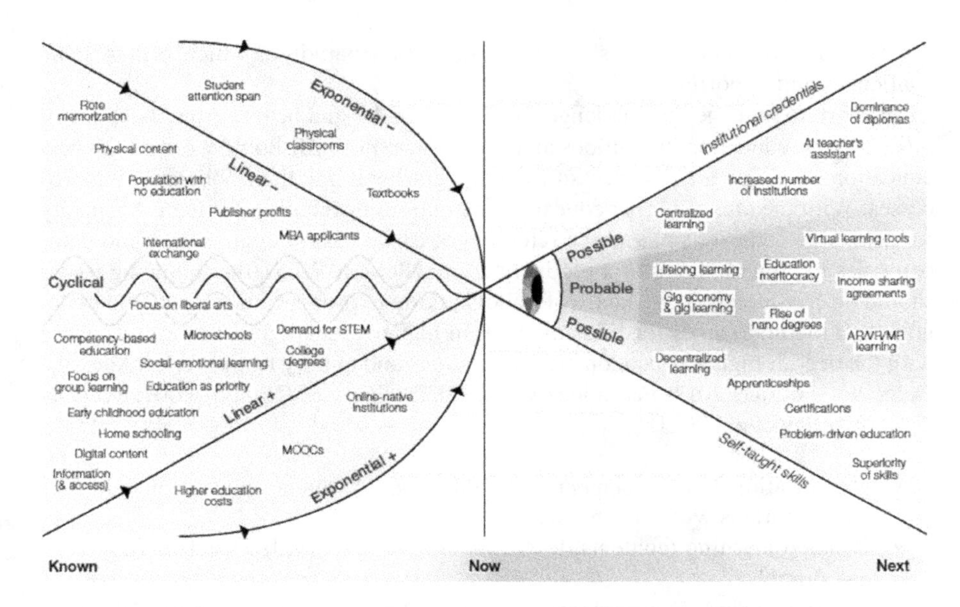

FIGURE 12.2 Education LEnS (longitudinal emergence scatter plot). Source: Taken from WEF (2021b) *Technology Futures: Projecting the Possible, Navigating What's Next*, p. 38.

expected to continue to increase, which include digital classes/labs, augmented reality, and virtual reality learning experiences.

As the report suggests, educational norms relating to when, where, and how people learn, and what they prefer to learn are waning, and this is expected to continue. It is not certain whether this will happen in the near future, as shown in the Education LEnS, but it is possible that traditional educational norms will continue to decline.

While teaching and learning environments are evolving with the concept of smart learning environments, they are making learning anywhere and any time possible with a variety of sub-concepts, such as flexible learning, personalised learning, mobile learning, adaptive learning, and blended learning (Cheung et al., 2021). They provide more student-centred approaches to content and delivery including evaluation and assessment stages. However, to design and deliver such content and tools through smart learning environments (Cheung et al., 2021), academics and higher education institution staff in general (including management and administrative staff) need to develop new competencies to build up such environments with new understandings and methodologies accordingly.

Meanwhile, academic/teacher identity, roles, and positions and the profession itself are another important topic of debate. In the studies, discussions take place on Education 4.0 and whether academics/teachers will lose their status as all-knowing experts, what the future classroom will look like, whether robots will replace human teachers, and so on. It is not easy to estimate these clearly; however, it is clear that a paradigm shift and a new epistemology are emerging. Furthermore, it is a shared responsibility to manage the shift in the education paradigm which brings some challenges and opportunities.

In short, to manage the challenges of Education 4.0 and benefit from its opportunities, higher education institutions and their academic staff should work on the new education paradigm and its necessities by integrating it into their policymaking processes. With the aim of higher education institutions developing graduates' capacity for academic achievement and the retention of certain knowledge, skills, and competencies to prepare them for a productive life (Gleason, 2018), the following issues should be considered at the centre of policymaking to produce an institutional pedagogy and methodologies for Education 4.0 in higher education institutions aligning with the overall higher education area regulations and quality frameworks (Adekola, Dale, & Gardiner, 2017; Fernández-Cruz, 2020; Guest & Clinton, 2007; Waring, 2017; Whalley et al., 2021):

- Organisational management and educational management (including collaborations with stakeholders)
- Blended learning (contents, tools, and delivery) and infrastructure (physical, digital)
- Smart learning environments (including data management, ethical/legal aspects)
- Continuous support services (including technology based, learning based, teaching based, counselling)
- Institutional culture and climate and supporting well-being
- Continuous professional development (for academics, managers/heads, administrative staff)

REFERENCES

Adekola, J., Dale, V.H., and Gardiner, K. (2017). Development of an institutional framework to guide transitions into enhanced blended learning in higher education. *Research in Learning Technology* 25:1–16. DOI: 10.25304/rlt.v25.1973

Bonfield, C.A., Salter, M., Longmuir, A., Benson, M., and Adachi, C. (2020) Transformation or evolution?: Education 4.0, teaching and learning in the digital age. *Higher Education Pedagogies* 5(1):223–246. DOI: 10.1080/23752696.2020.1816847

Cevik-Onar, S., Ustundag, A., Kadaifci, Ç., and Oztaysi, B. (2018) The changing role of engineering education in Industry 4.0 Era. In *Industry 4.0: Managing the Digital Transformation*, A. Ustundag, E. Cevikcan, eds. Springer Series in Advanced Manufacturing, pp. 137–151. Cham: Springer. DOI: 10.1007/978-3-319-57870-5_8

Cheung, S., Kwok, L.F., Phusavat, K., and Yang, H.H. (2021). Shaping the future learning environments with smart elements: Challenges and opportunities. *International Journal of Educational Technology in Higher Education* 18(1):16. DOI: 10.1186/s41239-021-00254-1

Echeberria, A.L. (2020). *A Digital Framework for Industry 4.0 Managing Strategy*. Cham: Palgrave Macmillan.

Fernández Cruz, M., Álvarez Rodríguez, J., and Ávalos Ruiz, I. (2020). Evaluation of the emotional and cognitive regulation of young people in a lockdown situation due to the Covid-19 pandemic. *Frontiers in Psychology* 11:565503. DOI: 10.3389/fpsyg.2020.565503

Framework for 21st Century Learning (2019). Accessed: 17 March 2020. www.battelleforkids.org/networks/p21/frameworks-resources

Gleason, N.W. (2018). *Higher Education in the Era of the Fourth Industrial Revolution*. Singapore: Palgrave Macmillan.

Guest, D. and Clinton, M. (2007). *Human Resource Management and University Performance*. Research and development series final report. London: Leadership Foundation for Higher Education.

Karaferye, F. (2020). Yükseköğretimde insan kaynağı. In *Yükseköğretim üzerine düşünmek*, ed. G.A. Baskan and N. Cemaloglu, 91–131. Ankara: Pegem Akademi, ISBN:978-625-7228-53-4.

Matthews, A., McLinden, M., and Greenway, C. (2021). Rising to the pedagogical challenges of the Fourth Industrial Age in the university of the future: An integrated model of scholarship. *Higher Education Pedagogies* 6(1):1–21. DOI: 10.1080/23752696.2020.1866440

OECD (2004). *On the Edge: Securing a Sustainable Future for Higher Education*. Organisation for Economic Co-Operation and Development.

OECD (2018). *The Future of Education and Skills Education 2030*. Organisation for Economic Co-Operation and Development.

OECD (2019). *The Survey of Adult Skills: Reader's Companion*, Third Edition, Paris: OECD Skills Studies, OECD Publishing. DOI: 10.1787/f70238c7-en

OECD (2020). *Back to the Future of Education: Four OECD Scenarios for Schooling*, Educational Research and Innovation. Paris: OECD Publishing. DOI: 10.1787/178ef527-en

PricewaterhouseCoopers (PwC) (2018). Will robots really steal our jobs? An international analysis of the potential long term impact of automation. Retrieved 19 April 2021 from www.pwc.com/hu/hu/kiadvanyok/assets/pdf/impact_of_automation_on_jobs.pdf

Raghunath, R., Anker, C., and Nortcliffe, A. (2018). Are academics ready for smart learning? *British Journal of Educational Technology* 49:182–197. DOI: 10.1111/bjet.12532

Sahu, P. (2020). Closure of universities due to coronavirus disease 2019 (COVID-19): Impact on education and mental health of students and academic staff. *Cureus* 12(4):e7541. DOI: 10.7759/cureus.7541

UNESCO (2014). *UNESCO Education Strategy 2014–2021*. Paris: The United Nations Educational, Scientific and Cultural Organization.

Xing, B. and Marwala, T. (2017). Implications of the fourth industrial age for higher education. *The Thinker* 73:10–15.

Xu, L.D., Xu, E.L., and Li, L. (2018). Industry 4.0: State of the art and future trends. *International Journal of Production Research* 56(8):2941–2962. DOI: 10.1080/00207543.2018.1444806

Waring, M. (2017). Management and leadership in UK universities: Exploring the possibilities of change. *Journal of Higher Education Policy and Management* 39(5):540–558. DOI: 10.1080/1360080X.2017.1354754

Whalley, B., France, D., Park, J., Mauchline, A., and Welsh, K. (2021). Towards flexible personalized learning and the future educational system in the fourth industrial revolution in the wake of Covid-19. *Higher Education Pedagogies* 6(1):79–99. DOI: 10.1080/23752696.2021.1883458

World Bank. (2019). *World Development Report 2019: The Changing Nature of Work*. Washington, DC: World Bank. DOI:10.1596/978-1-4648-1328-3

World Economic Forum [WEF] (2018a). *The Future of Jobs Report*. Retrieved from www3.weforum.org/docs/WEF_Future_of_Jobs_2018.pdf

World Economic Forum [WEF] (2018b). *Towards a Reskilling Revolution a Future of Jobs for All*. Retrieved from www3.weforum.org/docs/WEF_FOW_Reskilling_Revolution.pdf

World Economic Forum [WEF] (2020a). *The Future of Jobs Report 2020.* Retrieved from www.weforum.org/reports

World Economic Forum [WEF] (2020b). *Closing the Skills Gap: Key Insights and Success Metrics.* Retrieved from www3.weforum.org/docs/WEF_GSC_NES_White_Paper_2020.pdf

World Economic Forum [WEF] (2020c). *Schools of the Future Defining New Models of Education for the Fourth Industrial Revolution.* Retrieved from www.weforum.org/reports

World Economic Forum [WEF] (2021a). *Building a Common Language for Skills at Work: A Global Taxonomy.* Retrieved from www.weforum.org/reports

World Economic Forum [WEF] (2021b). *Technology Futures: Projecting the Possible, Navigating What's Next.* Retrieved from www.weforum.org/reports

Zhou, X. and Milecka-Forrest, M. (2021). Two groups separated by a shared goal: How academic managers and lecturers have embraced the introduction of digital technologies in UK Higher Education. *Research in Learning Technology* 29:1–18. DOI: 10.25304/rlt.v29.2446

Index

D

E

S

T

U

V

W

CPSIA information can be obtained
at www.ICGtesting.com
Printed in the USA
BVHW050050220522
637447BV00002B/10